Spring Boot+Vue
全栈开发实战

王松 著

内 容 简 介

Spring Boot 致力于简化开发配置并为企业级开发提供一系列非业务性功能,而 Vue 则采用数据驱动视图的方式将程序员从烦琐的 DOM 操作中解救出来。利用 Spring Boot+Vue,我们可以快速开发出大型 SPA 应用。

本书分为 16 章,重点讲解 Spring Boot 2 + Vue 2 全栈开发所涉及的各种技术点。所有技术点都配有操作实例,循序渐进,直到引导读者开发出一个完整的微人事 SPA 应用。

本书适合有一定基础的 Java 开发者及 Spring Boot 初学者学习,也适合高等院校和培训学校相关专业的师生作为教学参考书。

本书封面贴有清华大学出版社防伪标签,无标签者不得销售。
版权所有,侵权必究。举报:010-62782989,beiqinquan@tup.tsinghua.edu.cn。

图书在版编目(CIP)数据

Spring Boot+Vue 全栈开发实战/王松著.—北京:清华大学出版社,2019(2024.8重印)
ISBN 978-7-302-51797-9

Ⅰ. ①S… Ⅱ. ①王… Ⅲ. ①JAVA 语言—程序设计 Ⅳ. ①TP312.8

中国版本图书馆 CIP 数据核字(2018)第 269524 号

责任编辑:夏毓彦
封面设计:王 翔
责任校对:闫秀华
责任印制:曹婉颖

出版发行:清华大学出版社
网　　址:https://www.tup.com.cn,https://www.wqxuetang.com
地　　址:北京清华大学学研大厦 A 座　　　　　　邮　编:100084
社 总 机:010-83470000　　　　　　　　　　　　邮　购:010-62786544
投稿与读者服务:010-62776969,c-service@tup.tsinghua.edu.cn
质 量 反 馈:010-62772015,zhiliang@tup.tsinghua.edu.cn

印 装 者:三河市龙大印装有限公司
经　　销:全国新华书店
开　　本:190mm×260mm　　　　印　张:21.75　　　　字　数:557 千字
版　　次:2019 年 1 月第 1 版　　　　　　　　　　　印　次:2024 年 8 月第 20 次印刷
定　　价:69.00 元

产品编号:081014-01

前　　言

接触 Spring Boot 有好几年了，也曾断断续续出过一些教程，但是都比较零散，所使用的 Spring Boot 版本比较老，一直希望能够系统地写一本 Spring Boot 相关的图书，后来终于下定决心，在工作之余加班加点，于是有了读者现在所看到的这本书。

传统的 Spring 项目环境配置复杂臃肿，开发者早已不堪其苦，Spring Boot 带来的全新自动化配置解决方案一出现就受到了极大的关注，使得 Spring Boot 这两年成为 Java 领域的焦点之一。本书基于 Spring Boot 2.0.4（该版本是作者写作本书时 Spring Boot 的最新版本）完成。相对于 Spring Boot 1.5.X，Spring Boot 2 带来了许多新变化，这些在本书的相关章节都有体现。

本书分为 16 章，从以下方面向读者介绍 Spring Boot：

第 1 章　Spring Boot 入门
第 2 章　Spring Boot 基础配置
第 3 章　Spring Boot 整合视图层技术
第 4 章　Spring Boot 整合 Web 开发
第 5 章　Spring Boot 整合持久层技术
第 6 章　Spring Boot 整合 NoSQL
第 7 章　构建 RESTful 服务
第 8 章　开发者工具与单元测试
第 9 章　Spring Boot 缓存
第 10 章　Spring Boot 安全管理
第 11 章　Spring Boot 整合 WebSocket
第 12 章　消息服务
第 13 章　企业开发
第 14 章　应用监控
第 15 章　项目构建与部署
第 16 章　微人事项目实战

其中，第 1~15 章从视图层技术、持久化技术、NoSQL、RESTful、缓存、安全、WebSocket、消息服务以及企业开发等各个技术点对 Spring Boot 进行介绍；第 16 章通过一个 Spring Boot+Vue 搭建的前后端分离项目带领读者将前面 15 章所学的技术点应用到项目中，使读者深入体会前后端分离带来的好处，并学会搭建前后端分离的项目架构。

读者定位

本书适合有一定 Java Web 基础的开发者阅读，零基础的读者可以先学习 Java SE 和 Java Web 基础，再来阅读本书。

代码下载

本书示例源代码请扫描右边的二维码下载。如果下载有问题，请联系 booksaga@163.com，邮件主题为"Spring Boot+Vue 全栈开发实战"。

技术支持

由于水平有限，疏漏之处在所难免，若读者发现疏漏之处，可以通过以下方式联系作者：

- 邮箱：wangsong0210@gmail.com
- 微信：ws584991843
- 微信公众号：牧码小子

最后，祝每位读者阅读本书后都会有所收获，有所成长！

作者

2018 年 10 月

目　录

第 1 章　Spring Boot 入门 1
1.1　Spring Boot 简介 1
1.2　开发第一个 Spring Boot 程序 2
1.2.1　创建 Maven 工程 2
1.2.2　项目构建 5
1.2.3　项目启动 7
1.3　Spring Boot 的简便创建方式 8
1.3.1　在线创建 8
1.3.2　使用 IntelliJ IDEA 创建 9
1.3.3　使用 STS 创建 11
1.4　小结 12

第 2 章　Spring Boot 基础配置 13
2.1　不使用 spring-boot-starter-parent 13
2.2　@Spring BootApplication 14
2.3　定制 banner 16
2.4　Web 容器配置 17
2.4.1　Tomcat 配置 17
2.4.2　Jetty 配置 20
2.4.3　Undertow 配置 21
2.5　Properties 配置 21
2.6　类型安全配置属性 22
2.7　YAML 配置 24
2.7.1　常规配置 24
2.7.2　复杂配置 25
2.8　Profile 26
2.9　小结 27

第 3 章　Spring Boot 整合视图层技术 28
3.1　整合 Thymeleaf 28
3.2　整合 FreeMarker 31

3.3 小结 33

第 4 章 Spring Boot 整合 Web 开发 34

4.1 返回 JSON 数据 34
    4.1.1 默认实现 34
    4.1.2 自定义转换器 36
4.2 静态资源访问 40
    4.2.1 默认策略 40
    4.2.2 自定义策略 42
4.3 文件上传 42
    4.3.1 单文件上传 43
    4.3.2 多文件上传 45
4.4 @ControllerAdvice 46
    4.4.1 全局异常处理 46
    4.4.2 添加全局数据 48
    4.4.3 请求参数预处理 48
4.5 自定义错误页 50
    4.5.1 简单配置 52
    4.5.2 复杂配置 55
4.6 CORS 支持 62
4.7 配置类与 XML 配置 67
4.8 注册拦截器 68
4.9 启动系统任务 70
    4.9.1 CommandLineRunner 70
    4.9.2 ApplicationRunner 72
4.10 整合 Servlet、Filter 和 Listener 73
4.11 路径映射 75
4.12 配置 AOP 75
    4.12.1 AOP 简介 75
    4.12.2 Spring Boot 支持 76
4.13 其他 78
    4.13.1 自定义欢迎页 78
    4.13.2 自定义 favicon 79
    4.13.3 除去某个自动配置 79
4.14 小结 80

第 5 章 Spring Boot 整合持久层技术 81

5.1 整合 JdbcTemplate 81
5.2 整合 MyBatis 86

目录

- 5.3 整合 Spring Data JPA ... 89
- 5.4 多数据源 ... 95
  - 5.4.1 JdbcTemplate 多数据源 ... 96
  - 5.4.2 MyBatis 多数据源 ... 99
  - 5.4.3 JPA 多数据源 ... 102
- 5.5 小结 ... 106

第6章 Spring Boot 整合 NoSQL ... 107

- 6.1 整合 Redis ... 108
  - 6.1.1 Redis 简介 ... 108
  - 6.1.2 Redis 安装 ... 108
  - 6.1.3 整合 Spring Boot ... 110
  - 6.1.4 Redis 集群整合 Spring Boot ... 113
- 6.2 整合 MongoDB ... 121
  - 6.2.1 MongoDB 简介 ... 121
  - 6.2.2 MongoDB 安装 ... 122
  - 6.2.3 整合 Spring Boot ... 124
- 6.3 Session 共享 ... 127
  - 6.3.1 Session 共享配置 ... 128
  - 6.3.2 Nginx 负载均衡 ... 129
  - 6.3.3 请求分发 ... 130
- 6.4 小结 ... 131

第7章 构建 RESTful 服务 ... 132

- 7.1 REST 简介 ... 132
- 7.2 JPA 实现 REST ... 133
  - 7.2.1 基本实现 ... 133
  - 7.2.2 自定义请求路径 ... 138
  - 7.2.3 自定义查询方法 ... 138
  - 7.2.4 隐藏方法 ... 139
  - 7.2.5 配置 CORS ... 140
  - 7.2.6 其他配置 ... 140
- 7.3 MongoDB 实现 REST ... 141
- 7.4 小结 ... 142

第8章 开发者工具与单元测试 ... 143

- 8.1 devtools 简介 ... 143
- 8.2 devtools 实战 ... 143
  - 8.2.1 基本用法 ... 143

          8.2.2　基本原理 145
          8.2.3　自定义监控资源 145
          8.2.4　使用 LiveReload 146
          8.2.5　禁用自动重启 147
          8.2.6　全局配置 147
     8.3　单元测试 148
          8.3.1　基本用法 148
          8.3.2　Service 测试 149
          8.3.3　Controller 测试 149
          8.3.4　JSON 测试 152
     8.4　小结 153

第 9 章　Spring Boot 缓存 154

     9.1　Ehcache 2.x 缓存 155
     9.2　Redis 单机缓存 159
     9.3　Redis 集群缓存 160
          9.3.1　搭建 Redis 集群 161
          9.3.2　配置缓存 161
          9.3.3　使用缓存 162
     9.4　小结 164

第 10 章　Spring Boot 安全管理 165

     10.1　Spring Security 的基本配置 165
          10.1.1　基本用法 166
          10.1.2　配置用户名和密码 167
          10.1.3　基于内存的认证 167
          10.1.4　HttpSecurity 168
          10.1.5　登录表单详细配置 170
          10.1.6　注销登录配置 172
          10.1.7　多个 HttpSecurity 173
          10.1.8　密码加密 174
          10.1.9　方法安全 176
     10.2　基于数据库的认证 177
     10.3　高级配置 182
          10.3.1　角色继承 182
          10.3.2　动态配置权限 183
     10.4　OAuth 2 187
          10.4.1　OAuth 2 简介 187
          10.4.2　OAuth 2 角色 187

	10.4.3	OAuth 2 授权流程	188
	10.4.4	授权模式	188
	10.4.5	实践	189
10.5	Spring Boot 整合 Shiro	195
	10.5.1	Shiro 简介	195
	10.5.2	整合 Shiro	195
10.6	小结	200

第 11 章 Spring Boot 整合 WebSocket 201

11.1	为什么需要 WebSocket	201
11.2	WebSocket 简介	202
11.3	Spring Boot 整合 WebSocket	203
	11.3.1 消息群发	204
	11.3.2 消息点对点发送	208
11.4	小结	213

第 12 章 消息服务 214

12.1	JMS	214
	12.1.1 JMS 简介	214
	12.1.2 Spring Boot 整合 JMS	215
12.2	AMQP	218
	12.2.1 AMQP 简介	218
	12.2.2 Spring Boot 整合 AMQP	218
12.3	小结	228

第 13 章 企业开发 229

13.1	邮件发送	229
	13.1.1 发送前的准备	229
	13.1.2 发送	231
13.2	定时任务	239
	13.2.1 @Scheduled	239
	13.2.2 Quartz	240
13.3	批处理	243
	13.3.1 Spring Batch 简介	243
	13.3.2 整合 Spring Boot	243
13.4	Swagger 2	248
	13.4.1 Swagger 2 简介	248
	13.4.2 整合 Spring Boot	248
13.5	数据校验	252

13.5.1　普通校验 252
13.5.2　分组校验 254
13.5.3　校验注解 255
13.6　小结 256

第 14 章　应用监控 257

14.1　端点配置 257
　　14.1.1　开启端点 257
　　14.1.2　暴露端点 259
　　14.1.3　端点保护 261
　　14.1.4　端点响应缓存 261
　　14.1.5　路径映射 262
　　14.1.6　CORS 支持 262
　　14.1.7　健康信息 263
　　14.1.8　应用信息 267
14.2　监控信息可视化 272
14.3　邮件报警 275
14.4　小结 276

第 15 章　项目构建与部署 277

15.1　JAR 277
　　15.1.1　项目打包 277
　　15.1.2　项目运行 279
　　15.1.3　创建可依赖的 JAR 280
　　15.1.4　文件排除 281
15.2　WAR 283
15.3　小结 283

第 16 章　微人事项目实战 284

16.1　项目简介 284
16.2　技术架构 285
　　16.2.1　Vue 简介 285
　　16.2.2　Element 简介 285
　　16.2.3　其他 286
16.3　项目构建 286
　　16.3.1　前端项目构建 286
　　16.3.2　后端项目构建 287
　　16.3.3　数据模型设计 287
16.4　登录模块 293

  16.4.1 后端接口实现 293
  16.4.2 前端实现 302
16.5 动态加载用户菜单 308
  16.5.1 后端接口实现 308
  16.5.2 前端实现 310
16.6 员工资料模块 315
  16.6.1 后端接口实现 315
  16.6.2 前端实现 316
16.7 配置邮件发送 319
16.8 员工资料导出 322
  16.8.1 后端接口实现 322
  16.8.2 前端实现 325
16.9 员工资料导入 325
  16.9.1 后端接口实现 325
  16.9.2 前端实现 329
16.10 在线聊天 330
  16.10.1 后端接口实现 330
  16.10.2 前端实现 331
16.11 前端项目打包 334
16.12 小结 335

第 1 章

Spring Boot 入门

本章概要

- Spring Boot 简介
- 开发第一个 Spring Boot 程序
- Spring Boot 的简便创建方式

1.1 Spring Boot 简介

Spring 作为一个轻量级的容器，在 Java EE 开发中得到了广泛的应用，但是 Spring 的配置烦琐臃肿，在和各种第三方框架进行整合时代码量都非常大，并且整合的代码大多是重复的，为了使开发者能够快速上手 Spring，利用 Spring 框架快速搭建 Java EE 项目，Spring Boot 应运而生。

Spring Boot 带来了全新的自动化配置解决方案，使用 Spring Boot 可以快速创建基于 Spring 生产级的独立应用程序。Spring Boot 中对一些常用的第三方库提供了默认的自动化配置方案，使得开发者只需要很少的 Spring 配置就能运行一个完整的 Java EE 应用。Spring Boot 项目可以采用传统的方案打成 war 包，然后部署到 Tomcat 中运行。也可以直接打成可执行 jar 包，这样通过 java -jar 命令就可以启动一个 Spring Boot 项目。总体来说，Spring Boot 主要有如下优势：

- 提供一个快速的 Spring 项目搭建渠道。
- 开箱即用，很少的 Spring 配置就能运行一个 Java EE 项目。
- 提供了生产级的服务监控方案。

- 内嵌服务器，可以快速部署。
- 提供了一系列非功能性的通用配置。
- 纯 Java 配置，没有代码生成，也不需要 XML 配置。

Spring Boot 是一个"年轻"的项目，发展非常迅速，特别是在 Spring Boot 2.0 之后，许多 API 都有较大的变化，本书的写作基于目前最新的稳定版 2.0.4（本书写作时的最新版），因此需要 Java 8 或 9 以及 Spring Framework 5.0.8.RELEASE 或更高版本，同时，构建工具的版本要求为 Maven 3.2+ 或 Gradle 4。

1.2 开发第一个 Spring Boot 程序

Spring Boot 工程可以通过很多方式来创建，最通用的方式莫过于使用 Maven 了，因为大多数的 IDE 都支持 Maven。

1.2.1 创建 Maven 工程

这里不过多说明，Maven 的介绍和安装只介绍三种创建 Maven 工程的方式。

1. 使用命令创建 Maven 工程

首先可以通过 Maven 命令创建一个 Maven 工程，在 cmd 窗口中执行如下命令：

```
1  mvn archetype:generate -DgroupId=org.sang -DartifactId=chapter01
2   -DarchetypeArtifactId =maven-archetype-quickstart  -DinteractiveMode=false
```

命令解释：

- **-DgroupId**　组织 Id（项目包名）。
- **-DartifactId**　ArtifactId（项目名称或者模块名称）。
- **-DarchetypeArtifactId**　项目骨架。
- **-DinteractiveMode**　是否使用交互模式。

使用命令将项目创建好之后，直接用 Eclipse 或者 IntelliJ IDEA 打开即可。

2. 在 Eclipse 中创建 Maven 工程

大部分的 IDE 工具都可以直接创建 Maven 工程。在 Eclipse 中创建 Maven 工程的步骤如下：

步骤01　创建项目时选择 Maven Project，如图 1-1 所示。

第 1 章　Spring Boot 入门 | 3

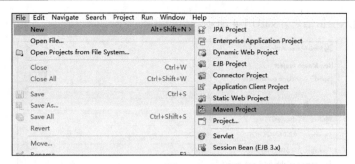

图 1-1

步骤02 选中 Use default Workspace location 复选框，如图 1-2 所示。

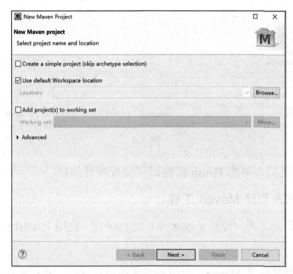

图 1-2

步骤03 选择项目骨架，保持默认设置即可，如图 1-3 所示。

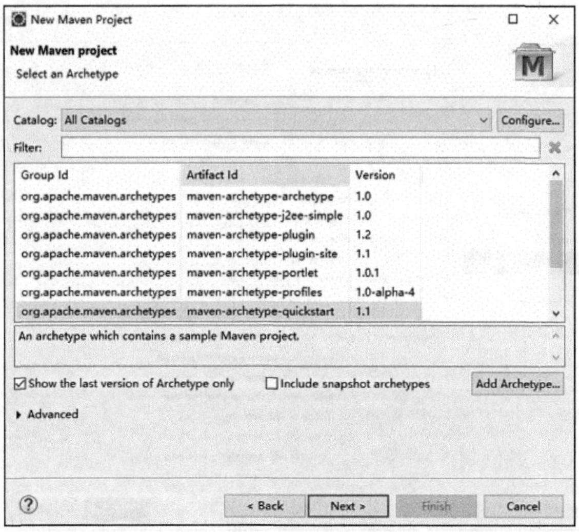

图 1-3

步骤04 输入项目信息，如图1-4所示。

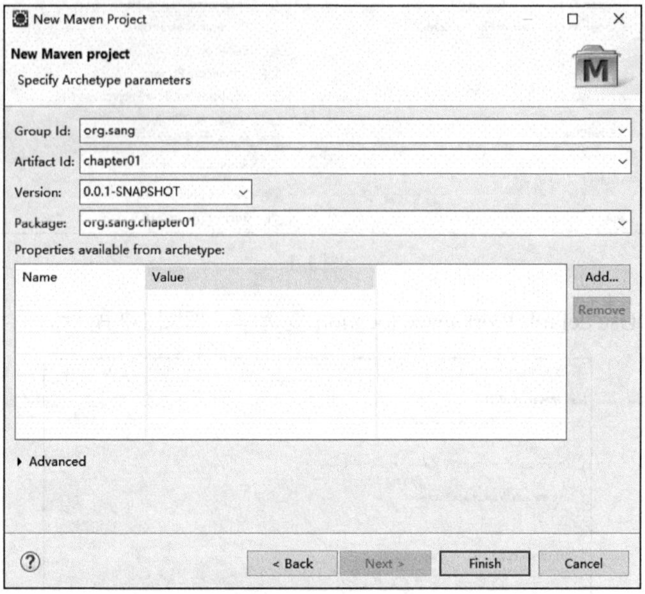

图1-4

完成以上4个步骤之后，单击Finish按钮即可完成项目创建。

3. 使用IntelliJ IDEA创建Maven工程

IntelliJ IDEA作为后起之秀，得到了越来越广泛的应用。使用IntelliJ IDEA创建Maven工程的步骤如下：

步骤01 创建项目时选择Maven，但是不必选择项目骨架，直接单击Next按钮即可，如图1-5所示。

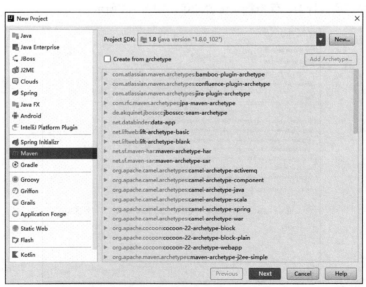

图1-5

步骤02 输入组织名称、模块名称、项目版本号等信息，如图 1-6 所示。

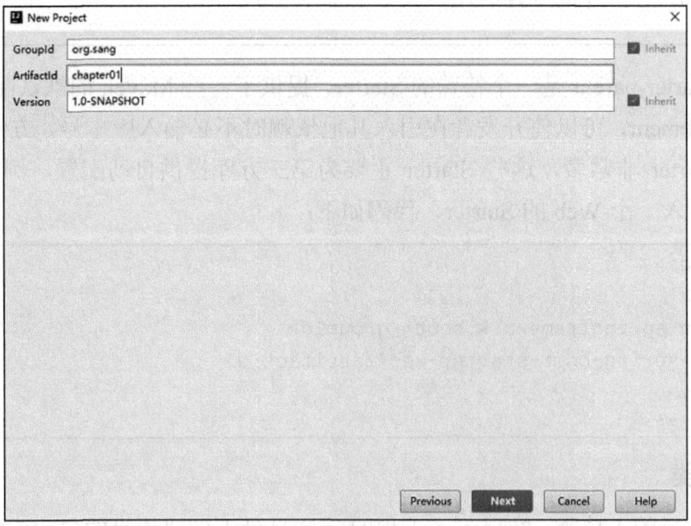

图 1-6

步骤03 选择项目位置，然后单击 Finish 按钮，完成项目创建，如图 1-7 所示。

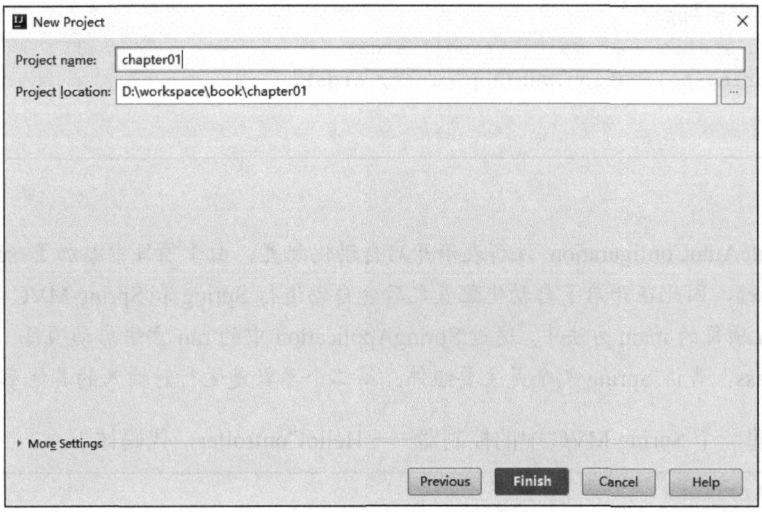

图 1-7

这里一共向读者介绍了三种创建 Maven 工程的方式，创建成功之后，接下来添加项目依赖。

1.2.2 项目构建

1. 添加依赖

首先添加 spring-boot-starter-parent 作为 parent，代码如下：

```
1  <parent>
2      <groupId>org.springframework.boot</groupId>
```

```
3    <artifactId>spring-boot-starter-parent</artifactId>
4    <version>2.0.4.RELEASE</version>
5    </parent>
```

spring-boot-starter-parent 是一个特殊的 Starter，提供了一些 Maven 的默认配置，同时还提供了 dependency-management，可以使开发者在引入其他依赖时不必输入版本号，方便依赖管理。Spring Boot 中提供的 Starter 非常多，这些 Starter 主要为第三方库提供自动配置，例如要开发一个 Web 项目，就可以先引入一个 Web 的 Starter，代码如下：

```
1    <dependencies>
2    <dependency>
3    <groupId>org.springframework.boot</groupId>
4    <artifactId>spring-boot-starter-web</artifactId>
5    </dependency>
6    </dependencies>
```

2. 编写启动类

接下来创建项目的入口类，在 Maven 工程的 java 目录下创建项目的包，包里创建一个 App 类，代码如下：

```
1    @EnableAutoConfiguration
2    public class App {
3        public static void main(String[] args) {
4            SpringApplication.run(App.class, args);
5        }
6    }
```

代码解释：

- @EnableAutoConfiguration 注解表示开启自动化配置。由于项目中添加了 spring-boot-starter-web 依赖，因此在开启了自动化配置之后会自动进行 Spring 和 Spring MVC 的配置。
- 在 Java 项目的 main 方法中，通过 SpringApplication 中的 run 方法启动项目。第一个参数传入 App.class，告诉 Spring 哪个是主要组件。第二个参数是运行时输入的其他参数。

接下来创建一个 Spring MVC 中的控制器——HelloController，代码如下：

```
1    @RestController
2    public class HelloController {
3        @GetMapping("/hello")
4        public String hello() {
5            return "hello spring boot!";
6        }
7    }
```

在控制器中提供了一个"/hello"接口，此时需要配置包扫描才能将 HelloController 注册到 Spring MVC 容器中，因此在 App 类上面再添加一个注解@ComponentScan 进行包扫描，代码如下：

```
1    @EnableAutoConfiguration
2    @ComponentScan
3    public class App {
```

```
4       public static void main(String[] args) {
5           SpringApplication.run(App.class, args);
6       }
7   }
```

也可以直接使用组合注解@Spring BootApplication 来代替@EnableAutoConfiguration 和@ComponentScan，代码如下：

```
1   @Spring BootApplication
2   public class App {
3       public static void main(String[] args) {
4           SpringApplication.run(App.class, args);
5       }
6   }
```

1.2.3 项目启动

启动项目有三种不同的方式，下面一一介绍。

1. 使用 Maven 命令启动

可以直接使用 mvn 命令启动项目，命令如下：

```
1   mvn spring-boot:run
```

启动成功后，在浏览器地址栏输入"http://localhost:8080/hello"即可看到运行结果，如图 1-8 所示。

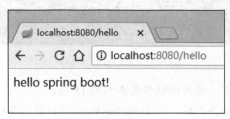

图 1-8

2. 直接运行 main 方法

直接在 IDE 中运行 App 类的 main 方法，就可以看到项目启动了，如图 1-9 所示。

```
  .   ____          _            __ _ _
 /\\ / ___'_ __ _ _(_)_ __  __ _ \ \ \ \
( ( )\___ | '_ | '_| | '_ \/ _` | \ \ \ \
 \\/  ___)| |_)| | | | | || (_| |  ) ) ) )
  '  |____| .__|_| |_|_| |_\__, | / / / /
 =========|_|==============|___/=/_/_/_/
 :: Spring Boot ::        (v2.0.3.RELEASE)
//其他日志
........o.s.b.w.embedded.tomcat.TomcatWebServer  : Tomcat started on port(s): 8080 (http) with context path ''
........org.sang.App                             : Started App in 7.876 seconds (JVM running for 9.518)
```

图 1-9

启动成功后，也可以在浏览器中直接访问/hello 接口。

3. 打包启动

当然,Spring Boot 应用也可以直接打成 jar 包运行。在生产环境中,也可以通过这样的方式来运行一个 Spring Boot 应用。要将 Spring Boot 打成 jar 包运行,首先需要添加一个 plugin 到 pom.xml 文件中,代码如下:

```
1  <build>
2    <plugins>
3      <plugin>
4        <groupId>org.springframework.boot</groupId>
5        <artifactId>spring-boot-maven-plugin</artifactId>
6      </plugin>
7    </plugins>
8  </build>
```

然后运行 mvn 命令进行打包,代码如下:

```
1  mvn package
```

打包完成后,在项目的 target 目录下会生成一个 jar 文件,通过 java -jar 命令直接启动这个 jar 文件,如图 1-10 所示。

图 1-10

关于打包启动的详细配置,读者可以参考本书第 15 章。

经过 1.2.1~1.2.3 小节的操作之后,一个 Spring Boot 项目就构建好并成功启动了。

1.3 Spring Boot 的简便创建方式

上面介绍的创建方式步骤有点多。在实际项目中,读者可结合具体的开发环境选择更简便的项目创建方式。下面介绍三种快捷创建方式。

1.3.1 在线创建

在线创建是 Spring Boot 官方提供的一种创建方式,在浏览器中输入网址 "https://start.spring.io/",可以看到如图 1-11 所示的页面。

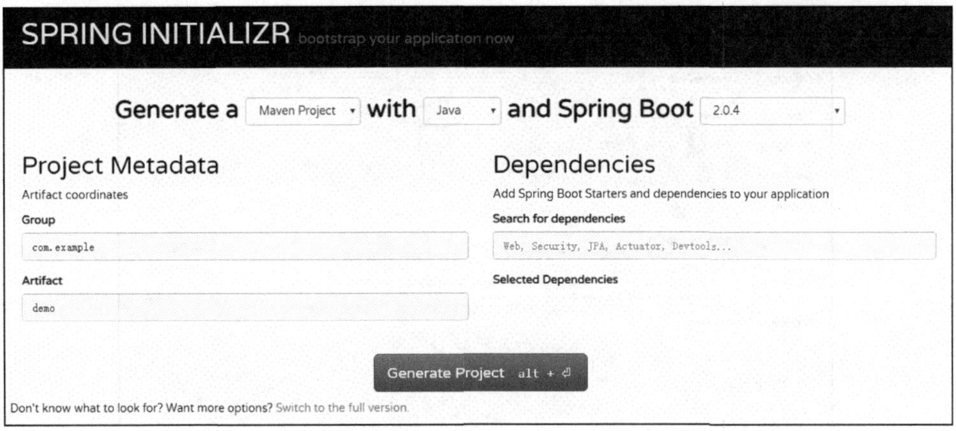

图 1-11

在这个页面中，可以选择项目的构建工具是 Maven 还是 Gradle、语言是 Java 还是其他、要使用的 Spring Boot 版本号、项目的组织 Id（包名）、模块名称以及项目的依赖。所有这些信息选好或填好后，单击 Generate Project 按钮将生成的模板下载到本地，解压后用 IDE 打开即可开始项目的开发。

1.3.2 使用 IntelliJ IDEA 创建

如果读者使用的开发工具是 IntelliJ IDEA，那么可以直接创建一个 Spring Boot 项目，但是注意直接创建 Spring Boot 项目这个功能在社区版的 IntelliJ IDEA 上是不存在的。创建方式如下：

步骤01 创建项目时选择 Spring Initializr，如图 1-12 所示。

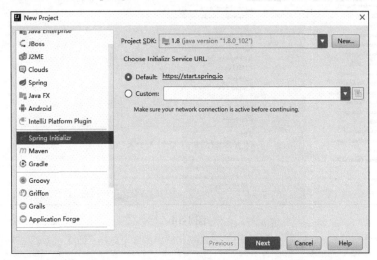

图 1-12

步骤02 输入项目基本信息，如图 1-13 所示。

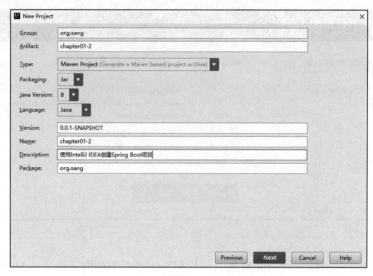

图 1-13

在这里输入项目的基本信息,包括组织 Id、模块名称、项目构建类型、最终生成包的类型、Java 的版本、开发语言、项目版本号、项目名称、项目描述以及项目的包。

步骤03 选择依赖,如图 1-14 所示。选择项目所需要添加的依赖,之后 IntelliJ IDEA 会自动把选中的依赖添加到项目的 pom.xml 文件中。

步骤04 选择项目创建路径,如图 1-15 所示。

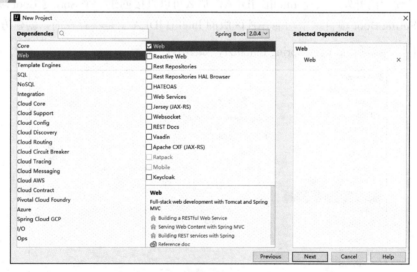

图 1-14

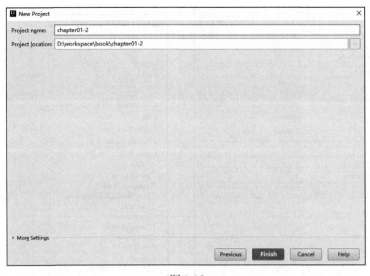

图 1-15

经过上面 4 个步骤之后，一个可运行的 Spring Boot 项目就创建成功了。本书后面的项目都将采用这种方式来创建。

1.3.3 使用 STS 创建

也有开发者习惯使用 STS 创建。在 STS 中创建 Spring Boot 项目也很方便：首先右击，选择 New→Spring Starter Project，如图 1-16 所示；然后在新页面中配置 Spring Boot 项目的基本信息（配置和 IntelliJ IDEA 的基本一致，不再赘述），如图 1-17 所示；最后选择需要添加的 Starter，选择完成后，单击 Finish 按钮完成项目创建，如图 1-18 所示。

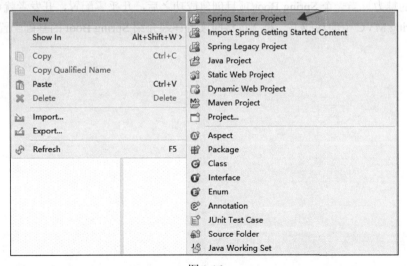

图 1-16

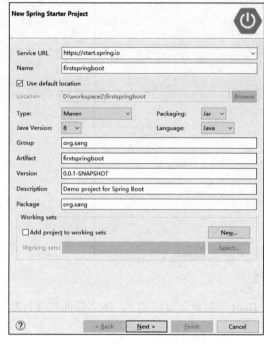
图 1-17

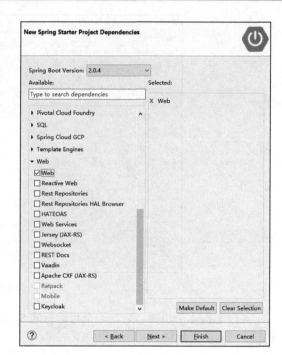
图 1-18

1.4 小　　结

本章主要向读者介绍了一个简单的 Spring Boot 项目的基本创建过程，从而让读者感受到 Spring Boot 的魅力。当一个 Spring Boot 项目创建成功之后，几乎零配置，开发者就可以直接使用 Spring 和 Spring MVC 中的功能了。第 2 章将向读者详细介绍 Spring Boot 的基础配置。

第 2 章

Spring Boot 基础配置

本章概要

- 不使用 spring-boot-starter-parent
- @Spring BootApplication
- 定制 banner
- Web 容器配置
- Properties 配置
- 类型安全配置属性
- YAML 配置
- Profile

2.1 不使用 spring-boot-starter-parent

从第 1 章的介绍中读者了解到在向 pom.xml 文件中添加依赖之前需要先添加 spring-boot-starter-parent。spring-boot-starter-parent 主要提供了如下默认配置：

- Java 版本默认使用 1.8。
- 编码格式默认使用 UTF-8。
- 提供 Dependency Management 进行项目依赖的版本管理。
- 默认的资源过滤与插件配置。

spring-boot-starter-parent 虽然方便，但是读者在公司中开发微服务项目或者多模块项目时一般需要使用公司自己的 parent，这个时候如果还想进行项目依赖版本的统一管理，就需要使用 dependencyManagement 来实现了。添加如下代码到 pom.xml 文件中：

```
1  <dependencyManagement>
2  <dependencies>
3  <dependency>
4  <groupId>org.springframework.boot</groupId>
5  <artifactId>spring-boot-dependencies</artifactId>
6  <version>2.0.4.RELEASE</version>
7  <type>pom</type>
8  <scope>import</scope>
9  </dependency>
10 </dependencies>
11 </dependencyManagement>
```

此时，就可以不用继承 spring-boot-starter-parent 了，但是 Java 的版本、编码的格式等都需要开发者手动配置。Java 版本的配置很简单，添加一个 plugin 即可：

```
1  <plugin>
2  <groupId>org.apache.maven.plugins</groupId>
3  <artifactId>maven-compiler-plugin</artifactId>
4  <version>3.1</version>
5  <configuration>
6  <source>1.8</source>
7  <target>1.8</target>
8  </configuration>
9  </plugin>
```

至于编码格式，如果采用了 1.3 节介绍的方式创建 Spring Boot 项目，那么编码格式默认会加上；如果是通过普通 Maven 项目配置成的 Spring Boot 项目，那么在 pom.xml 文件中加入如下配置即可：

```
1  <properties>
2  <project.build.sourceEncoding>UTF-8</project.build.sourceEncoding>
3  <project.reporting.outputEncoding>UTF-8</project.reporting.outputEncoding>
4  </properties>
```

2.2 @Spring BootApplication

在前文的介绍中，读者已经了解到@Spring BootApplication 注解是加在项目的启动类上的。@Spring BootApplication 实际上是一个组合注解，定义如下：

```
1  @Spring BootConfiguration
2  @EnableAutoConfiguration
3  @ComponentScan(excludeFilters = {
4      @Filter(type = FilterType.CUSTOM, classes = TypeExcludeFilter.class),
```

```
5       @Filter(type = FilterType.CUSTOM, classes =
6   AutoConfigurationExcludeFilter.class) })
7   public @interface Spring BootApplication {
8   //略
    }
```

这个注解由三个注解组成。

①第一个@Spring BootConfiguration 的定义如下：

```
1   @Configuration
2   public @interface Spring BootConfiguration {
3
4   }
```

原来就是一个@Configuration，所以@Spring BootConfiguration 的功能就是表明这是一个配置类，开发者可以在这个类中配置 Bean。从这个角度来讲，这个类所扮演的角色有点类似于 Spring 中 applicationContext.xml 文件的角色。

②第二个注解@EnableAutoConfiguration 表示开启自动化配置。Spring Boot 中的自动化配置是非侵入式的，在任意时刻，开发者都可以使用自定义配置代替自动化配置中的某一个配置。

③第三个注解@ComponentScan 完成包扫描，也是 Spring 中的功能。由于@ComponentScan 注解默认扫描的类都位于当前类所在包的下面，因此建议在实际项目开发中把项目启动类放在根包中，如图 2-1 所示。

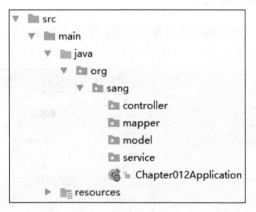

图 2-1

虽然项目的启动类也包含@Configuration 注解，但是开发者可以创建一个新的类专门用来配置 Bean，这样便于配置的管理。这个类只需要加上@Configuration 注解即可，代码如下：

```
1   @Configuration
2   public class MyConfig {
3   }
```

项目启动类中的@ComponentScan 注解，除了扫描@Service、@Repository、@Component、@Controller 和@RestController 等之外，也会扫描@Configuration 注解的类。

2.3 定制 banner

Spring Boot 项目在启动时会打印一个 banner，如图 2-2 所示。

图 2-2

这个 banner 是可以定制的，在 resources 目录下创建一个 banner.txt 文件，在这个文件中写入的文本将在项目启动时打印出来。如果想将 TXT 文本设置成艺术字体，有以下几个在线网站可供参考：

- http://www.network-science.de/ascii/
- http://www.kammerl.de/ascii/AsciiSignature.php
- http://patorjk.com/software/taag

以第一个网站为例，打开后输入要设置的文本，单击"do it！"按钮，将生成的文本复制到 banner.txt 文件中，如图 2-3 所示。

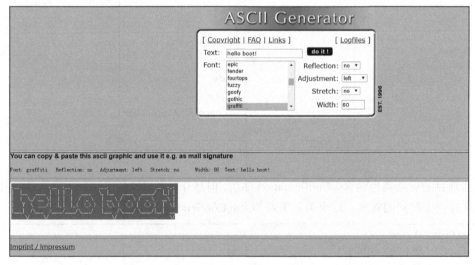

图 2-3

复制完成后再启动项目，就可以看到 banner 发生了变化，如图 2-4 所示。

```
  .   ____          _            __ _ _
 /\\ / ___'_ __ _ _(_)_ __  __ _ \ \ \ \
( ( )\___ | '_ | '_| | '_ \/ _` | \ \ \ \
 \\/  ___)| |_)| | | | | || (_| |  ) ) ) )
  '  |____| .__|_| |_|_| |_\__, | / / / /
 =========|_|==============|___/=/_/_/_/
2018-07-04 16:12:18.755  INFO 8968 --- [           main] org.sang.Chapter012Application
2018-07-04 16:12:18.761  INFO 8968 --- [           main] org.sang.Chapter012Application
```

图 2-4

想关闭 banner 也是可以的，修改项目启动类的 main 方法，代码如下：

```
1  public static void main(String[] args) {
2      SpringApplicationBuilder builder = new
3  SpringApplicationBuilder(Chapter012Application.class);
4      builder.bannerMode(Banner.Mode.OFF).run(args);
5  }
```

通过 SpringApplicationBuilder 来设置 bannerMode 为 OFF，这样启动时 banner 就消失了。

2.4 Web 容器配置

2.4.1 Tomcat 配置

1. 常规配置

在 Spring Boot 项目中，可以内置 Tomcat、Jetty、Undertow、Netty 等容器。当开发者添加了 spring-boot-starter-web 依赖之后，默认会使用 Tomcat 作为 Web 容器。如果需要对 Tomcat 做进一步的配置，可以在 application.properties 中进行配置，代码如下：

```
1  server.port=8081
2  server.error.path=/error
3  server.servlet.session.timeout=30m
4  server.servlet.context-path=/chapter02
5  server.tomcat.uri-encoding=utf-8
6  server.tomcat.max-threads=500
7  server.tomcat.basedir=/home/sang/tmp
```

代码解释：

- server.port 配置了 Web 容器的端口号。
- error.path 配置了当项目出错时跳转去的页面。
- session.timeout 配置了 session 失效时间，30m 表示 30 分钟，如果不写单位，默认单位是秒。由于 Tomcat 中配置 session 过期时间以分钟为单位，因此这里单位如果是秒的话，该时间会被转换为一个不超过所配置秒数的最大分钟数，例如这里配置了 119，默认单位为秒，则实际 session 过期时间为 1 分钟。
- context-path 表示项目名称，不配置时默认为/。如果配置了，就要在访问路径中加上配置的路径。
- uri-encoding 表示配置 Tomcat 请求编码。
- max-threads 表示 Tomcat 最大线程数。
- basedir 是一个存放 Tomcat 运行日志和临时文件的目录，若不配置，则默认使用系统的临时目录。

当然，Web 容器相关的配置不止这些，这里只列出了一些常用的配置，完整的配置可以参考官方文档 Appendix A. Common application properties 一节。

2. HTTPS 配置

由于 HTTPS 具有良好的安全性，在开发中得到了越来越广泛的应用，像微信公众号、小程序等的开发都要使用 HTTPS 来完成。对于个人开发者而言，一个 HTTPS 证书的价格还是有点贵，国内有一些云服务器厂商提供免费的 HTTPS 证书，一个账号可以申请数个。不过在 jdk 中提供了一个 Java 数字证书管理工具 keytool，在\jdk\bin 目录下，通过这个工具可以自己生成一个数字证书，生成命令如下：

```
1  keytool -genkey -alias tomcathttps -keyalg RSA -keysize 2048 -keystore sang.p12 -validity 365
```

命令解释：
- -genkey 表示要创建一个新的密钥。
- -alias 表示 keystore 的别名。
- -keyalg 表示使用的加密算法是 RSA，一种非对称加密算法。
- -keysize 表示密钥的长度。
- -keystore 表示生成的密钥存放位置。
- -validity 表示密钥的有效时间，单位为天。

在 cmd 窗口中直接执行如上命令，在执行的过程中需要输入密钥口令等信息，根据提示输入即可。命令执行完成后，会在当前用户目录下生成一个名为 sang.p12 的文件，将这个文件复制到项目的根目录下，然后在 application.properties 中做如下配置：

```
1  server.ssl.key-store=sang.p12
2  server.ssl.key-alias=tomcathttps
3  server.ssl.key-store-password=123456
```

代码解释：
- key-store 表示密钥文件名。
- key-alias 表示密钥别名。
- key-store-password 就是在 cmd 命令执行过程中输入的密码。

配置成功后，启动项目，在浏览器中输入"https://localhost:8081/chapter02/hello"来查看结果。注意，证书是自己生成的，不被浏览器认可，此时添加信任或者继续前进即可，如图 2-5 所示。

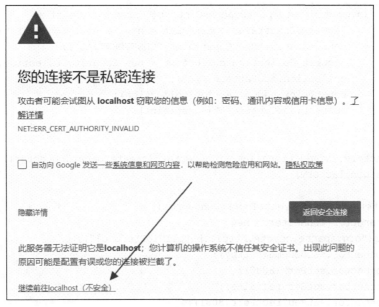

图 2-5

成功运行的结果如图 2-6 所示。

图 2-6

此时,如果以 HTTP 的方式访问接口,就会访问失败,如图 2-7 所示。

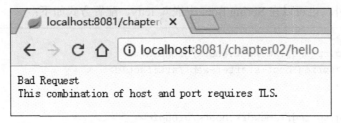

图 2-7

这是因为 Spring Boot 不支持同时在配置中启动 HTTP 和 HTTPS。这个时候可以配置请求重定向,将 HTTP 请求重定向为 HTTPS 请求。配置方式如下:

```
1  @Configuration
2  public class TomcatConfig {
3      @Bean
4      TomcatServletWebServerFactory tomcatServletWebServerFactory() {
5          TomcatServletWebServerFactory factory = new TomcatServletWebServerFactory(){
6              @Override
```

```
7             protected void postProcessContext(Context context) {
8                 SecurityConstraint constraint = new SecurityConstraint();
9                 constraint.setUserConstraint("CONFIDENTIAL");
10                SecurityCollection collection = new SecurityCollection();
11                collection.addPattern("/*");
12                constraint.addCollection(collection);
13                context.addConstraint(constraint);
14            }
15        };
16        factory.addAdditionalTomcatConnectors(createTomcatConnector());
17        return factory;
18    }
19    private Connector createTomcatConnector() {
20        Connector connector = new
21 Connector("org.apache.coyote.http11.Http11NioProtocol");
22        connector.setScheme("http");
23        connector.setPort(8080);
24        connector.setSecure(false);
25        connector.setRedirectPort(8081);
26        return connector;
27    }
28 }
```

这里首先配置一个 TomcatServletWebServerFactory，然后添加一个 Tomcat 中的 Connector（监听 8080 端口），并将请求转发到 8081 上去。

配置完成后，在浏览器中输入"http://localhost:8080/chapter02/hello"，就会自动重定向到 https://localhost:8081/chapter02/hello 上。

2.4.2 Jetty 配置

除了 Tomcat 外，也可以在 Spring Boot 中嵌入 Jetty，配置方式如下：

```
1  <dependency>
2      <groupId>org.springframework.boot</groupId>
3      <artifactId>spring-boot-starter-web</artifactId>
4      <exclusions>
5          <exclusion>
6              <groupId>org.springframework.boot</groupId>
7              <artifactId>spring-boot-starter-tomcat</artifactId>
8          </exclusion>
9      </exclusions>
10 </dependency>
11 <dependency>
12     <groupId>org.springframework.boot</groupId>
13     <artifactId>spring-boot-starter-jetty</artifactId>
14 </dependency>
```

主要是从 spring-boot-starter-web 中除去默认的 Tomcat，然后加入 Jetty 的依赖即可。此时启动项目，查看启动日志，如图 2-8 所示。

```
FrameworkServlet 'dispatcherServlet': initialization completed in 17 ms
Started ServerConnector@2617f816{HTTP/1.1, [http/1.1]}{0.0.0.0:8080}
Jetty started on port(s) 8080 (http/1.1) with context path '/'
Started Chapter013Application in 8.447 seconds (JVM running for 9.865)
```

图 2-8

2.4.3 Undertow 配置

Undertow 是一个红帽公司开源的 Java 服务器，具有非常好的性能，在 Spring Boot 中也得到了很好的支持，配置方式与 Jetty 类似，代码如下：

```
1   <dependency>
2       <groupId>org.springframework.boot</groupId>
3       <artifactId>spring-boot-starter-web</artifactId>
4       <exclusions>
5           <exclusion>
6               <groupId>org.springframework.boot</groupId>
7               <artifactId>spring-boot-starter-tomcat</artifactId>
8           </exclusion>
9       </exclusions>
10  </dependency>
11  <dependency>
12      <groupId>org.springframework.boot</groupId>
13      <artifactId>spring-boot-starter-undertow</artifactId>
14  </dependency>
```

启动后查看日志，如图 2-9 所示。

```
Mapped URL path [/**] onto handler of type [class org.springframework.we
Registering beans for JMX exposure on startup
Undertow started on port(s) 8080 (http) with context path ''
Started Chapter013Application in 6.998 seconds (JVM running for 8.749)
```

图 2-9

2.5 Properties 配置

Spring Boot 中采用了大量的自动化配置，但是对开发者而言，在实际项目中不可避免会有一些需要自己手动配置，承载这些自定义配置的文件就是 resources 目录下的 application.properties 文件（也可以使用 YAML 配置来替代 application.properties 配置，YAML 配置将在 2.7 节介绍）。在 2.4 节的 Web 容器配置中，读者已经见识到 application.properties 配置的基本用法了，本节将对 application.properties 的使用做进一步的介绍。

Spring Boot 项目中的 application.properties 配置文件一共可以出现在如下 4 个位置：

- 项目根目录下的 config 文件夹中。
- 项目根目录下。
- classpath 下的 config 文件夹中。
- classpath 下。

如果这 4 个位置中都有 application.properties 文件，那么加载的优先级从 1 到 4 依次降低，如图 2-10 所示。Spring Boot 将按照这个优先级查找配置信息，并加载到 Spring Environment 中。

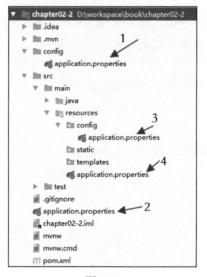

图 2-10

如果开发者在开发中未使用 application.properties，而是使用了 application.yml 作为配置文件，那么配置文件的优先级与图 2-10 一致。

默认情况下，Spring Boot 按照图 2-10 的顺序依次查找 application.properties 并加载。如果开发者不想使用 application.properties 作为配置文件名，也可以自己定义。例如，在 resources 目录下创建一个配置文件 app.properties，然后将项目打成 jar 包，打包成功后，使用如下命令运行：

```
1  java -jar chapter02-2-0.0.1-SNAPSHOT.jar --spring.config.name=app
```

在运行时再指定配置文件的名字。使用 spring.config.location 可以指定配置文件所在目录（注意需要以/结束），代码如下：

```
1  java -jar chapter02-2-0.0.1-SNAPSHOT.jar --spring.config.name=app
   --spring.config.location=classpath:/
```

2.6 类型安全配置属性

在 2.5 节中，读者已经了解到无论是 Properties 配置还是 YAML 配置，最终都会被加载到 Spring

Environment 中。Spring 提供了@Value 注解以及 EnvironmentAware 接口来将 Spring Environment 中的数据注入到属性上，Spring Boot 对此进一步提出了类型安全配置属性（Type-safe Configuration Properties），这样即使在数据量非常庞大的情况下，也可以更加方便地将配置文件中的数据注入 Bean 中。考虑在 application.properties 中添加如下一段配置：

```
1  book.name=三国演义
2  book.author=罗贯中
3  book.price=30
```

将这一段配置数据注入如下 Bean 中：

```
1  @Component
2  @ConfigurationProperties(prefix = "book")
3  public class Book {
4      private String name;
5      private String author;
6      private Float price;
7      //省略 getter/setter
8  }
```

代码解释：

- @ConfigurationProperties 中的 prefix 属性描述了要加载的配置文件的前缀。
- 如果配置文件是一个 YAML 文件，那么可以将数据注入一个集合中。YAML 将在 2.7 节介绍。
- Spring Boot 采用了一种宽松的规则来进行属性绑定，如果 Bean 中的属性名为 authorName，那么配置文件中的属性可以是 book.author_name、book.author-name、book.authorName 或者 book.AUTHORNAME。

> **注　意**
>
> 以上的配置可能会乱码，需要对中文进行转码。在 IntelliJ IDEA 中，这个转码非常容易，在 setting 配置中进行简单配置即可，如图 2-11 所示。

最后创建 BookController 进行简单测试：

```
1  @RestController
2  public class BookController {
3      @Autowired
4      Book book;
5      @GetMapping("/book")
6      public String book() {
7          return book.toString();
8      }
9  }
```

注入 Book，并将实例输出，如图 2-12 所示。

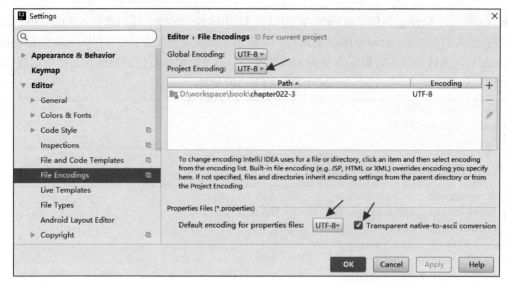

图 2-11

图 2-12

2.7 YAML 配置

2.7.1 常规配置

YAML 是 JSON 的超集，简洁而强大，是一种专门用来书写配置文件的语言，可以替代 application.properties。在创建一个 Spring Boot 项目时，引入的 spring-boot-starter-web 依赖间接地引入了 snakeyaml 依赖，snakeyaml 会实现对 YAML 配置的解析。YAML 的使用非常简单，利用缩进来表示层级关系，并且大小写敏感。在 Spring Boot 项目中使用 YAML 只需要在 resources 目录下创建一个 application.yml 文件即可，然后向 application.yml 中添加如下配置：

```
server:
  port: 80
  servlet:
    context-path: /chapter02
  tomcat:
    uri-encoding: utf-8
```

这一段配置等效于 application.properties 中的如下配置：

```
1  server.port=80
2  server.servlet.context-path=/chapter02
3  server.tomcat.uri-encoding=utf-8
```

此时可以将 resources 目录下的 application.properties 文件删除，完全使用 YAML 完成文件的配置。

2.7.2 复杂配置

YAML 不仅可以配置常规属性，也可以配置复杂属性，例如下面一组配置：

```
1  my:
2    name: 江南一点雨
3    address: China
```

像 Properties 配置文件一样，这一段配置也可以注入一个 Bean 中，代码如下：

```
1  @Component
2  @ConfigurationProperties(prefix = "my")
3  public class User {
4      private String name;
5      private String address;
6  //省略 getter/setter
7  }
```

YAML 还支持列表的配置，例如下面一组配置：

```
1  my:
2    name: 江南一点雨
3    address: China
4    favorites:
5      - 足球
6      - 徒步
7      - Coding
```

这一组配置可以注入如下 Bean 中：

```
1  @Component
2  @ConfigurationProperties(prefix = "my")
3  public class User {
4      private String name;
5      private String address;
6      private List<String> favorites;
7  //省略 getter/setter
8  }
```

YAML 还支持更复杂的配置，即集合中也可以是一个对象，例如下面一组配置：

```
1  my:
2    users:
3      - name: 江南一点雨
4        address: China
```

```
5          favorites:
6            - 足球
7            - 徒步
8            - Coding
9        - name: sang
10         address: GZ
11         favorites:
12           - 阅读
13           - 吉他
```

这组配置在集合中放的是一个对象，因此可以注入如下集合中：

```
1   @Component
2   @ConfigurationProperties(prefix = "my")
3   public class Users {
4       private List<User> users;
5   //省略getter/setter
6   }
7   public class User {
8       private String name;
9       private String address;
10      private List<String> favorites;
11  //省略getter/setter
12  }
```

在 Spring Boot 中使用 YAML 虽然方便，但是 YAML 也有一些缺陷，例如无法使用 @PropertySource 注解加载 YAML 文件，如果项目中有这种需求，还是需要使用 Properties 格式的配置文件。

2.8 Profile

开发者在项目发布之前，一般需要频繁地在开发环境、测试环境以及生产环境之间进行切换，这个时候大量的配置需要频繁更改，例如数据库配置、redis 配置、mongodb 配置、jms 配置等。频繁修改带来了巨大的工作量，Spring 对此提供了解决方案（@Profile 注解），Spring Boot 则更进一步提供了更加简洁的解决方案，Spring Boot 中约定的不同环境下配置文件名称规则为 application-{profile}.properties，profile 占位符表示当前环境的名称，具体配置步骤如下：

1. 创建配置文件

首先在 resources 目录下创建两个配置文件：application-dev.properties 和 application-prod.properties，分别表示开发环境中的配置和生产环境中的配置。其中，application-dev.properties 文件的内容如下：

```
1   server.port=8080
```

application-prod.properties 文件的内容如下：

```
1  server.port=80
```

这里为了简化问题并且容易看到效果，两个配置文件中主要修改了一下项目端口号。

2. 配置 application.properties

然后在 application.properties 中进行配置：

```
1  spring.profiles.active=dev
```

这个表示使用 application-dev.properties 配置文件启动项目，若将 dev 改为 prod，则表示使用 application-prod.properties 启动项目。项目启动成功后，就可以通过相应的端口进行访问了。

3. 在代码中配置

对于第二步在 application.properties 中添加的配置，我们也可以在代码中添加配置来完成，在启动类的 main 方法上添加如下代码，可以替换第二步的配置：

```
1  SpringApplicationBuilder builder = new
2  SpringApplicationBuilder(Chapter013Application.class);
3  builder.application().setAdditionalProfiles("prod");
4  builder.run(args);
```

4. 项目启动时配置

对于第 2 步和第 3 步提到的两种配置方式，也可以在将项目打成 jar 包后启动时，在命令行动态指定当前环境，示例命令如下：

```
1  java -jar chapter01-3-0.0.1-SNAPSHOT.jar --spring.profiles.active=prod
```

2.9 小　　结

本章主要向读者介绍了 Spring Boot 常见的基础性配置，包括依赖管理的多种方式，如入口类注解、banner 定制、Web 容器配置以及 Properties 配置和 YAML 配置等，这些配置将是后面章节的基础。第 3 章将向读者介绍使用 Spring Boot 整合视图层技术。

第 3 章

Spring Boot 整合视图层技术

本章概要

- 整合 Thymeleaf
- 整合 FreeMarker

在目前的企业级应用开发中，前后端分离是趋势，但是视图层技术还占有一席之地。Spring Boot 对视图层技术提供了很好的支持，官方推荐使用的模板引擎是 Thymeleaf，不过像 FreeMarker 也支持，JSP 技术在这里并不推荐使用。下面分别向读者介绍 Spring Boot 整合 Thymeleaf 和 FreeMarker 两种视图层技术。

3.1 整合 Thymeleaf

Thymeleaf 是新一代 Java 模板引擎，类似于 Velocity、FreeMarker 等传统 Java 模板引擎。与传统 Java 模板引擎不同的是，Thymeleaf 支持 HTML 原型，既可以让前端工程师在浏览器中直接打开查看样式，也可以让后端工程师结合真实数据查看显示效果。同时，Spring Boot 提供了 Thymeleaf 自动化配置解决方案，因此在 Spring Boot 中使用 Thymeleaf 非常方便。Spring Boot 整合 Thymeleaf 主要可通过如下步骤：

1. 创建工程，添加依赖

新建一个 Spring Boot 工程，然后添加 spring-boot-starter-web 和 spring-boot-starter-thymeleaf 依赖，代码如下：

```
1  <dependency>
2  <groupId>org.springframework.boot</groupId>
3  <artifactId>spring-boot-starter-thymeleaf</artifactId>
4  </dependency>
5  <dependency>
6  <groupId>org.springframework.boot</groupId>
7  <artifactId>spring-boot-starter-web</artifactId>
8  </dependency>
```

2. 配置 Thymeleaf

Spring Boot 为 Thymeleaf 提供了自动化配置类 ThymeleafAutoConfiguration，相关的配置属性在 ThymeleafProperties 类中，ThymeleafProperties 部分源码如下：

```
1  @ConfigurationProperties(prefix = "spring.thymeleaf")
2  public class ThymeleafProperties {
3      private static final Charset DEFAULT_ENCODING = StandardCharsets.UTF_8;
4      public static final String DEFAULT_PREFIX = "classpath:/templates/";
5      public static final String DEFAULT_SUFFIX = ".html";
6      …
7      …
8  }
```

由此配置可以看到，默认的模板位置在 classpath:/templates/，默认的模板后缀为.html。如果使用 IntelliJ IDEA 工具创建 Spring Boot 项目，templates 文件夹默认就会创建。

当然，如果开发者想对默认的 Thymeleaf 配置参数进行自定义配置，那么可以直接在 application.properties 中进行配置，部分常见配置如下：

```
1  #是否开启缓存，开发时可设置为 false，默认为 true
2  spring.thymeleaf.cache=true
3  #检查模板是否存在，默认为 true
4  spring.thymeleaf.check-template=true
5  #检查模板位置是否存在，默认为 true
6  spring.thymeleaf.check-template-location=true
7  #模板文件编码
8  spring.thymeleaf.encoding=UTF-8
9  #模板文件位置
10 spring.thymeleaf.prefix=classpath:/templates/
11 #Content-Type 配置
12 spring.thymeleaf.servlet.content-type=text/html
13 #模板文件后缀
14 spring.thymeleaf.suffix=.html
```

3. 配置控制器

创建 Book 实体类，然后在 Controller 中返回 ModelAndView，代码如下：

```
1  public class Book {
2      private Integer id;
3      private String name;
4      private String author;
5      //省略 getter/setter
```

```
6  }
7  @Controller
8  public class BookController {
9      @GetMapping("/books")
10     public ModelAndView books() {
11         List<Book> books = new ArrayList<>();
12         Book b1 = new Book();
13         b1.setId(1);
14         b1.setAuthor("罗贯中");
15         b1.setName("三国演义");
16         Book b2 = new Book();
17         b2.setId(2);
18         b2.setAuthor("曹雪芹");
19         b2.setName("红楼梦");
20         books.add(b1);
21         books.add(b2);
22         ModelAndView mv = new ModelAndView();
23         mv.addObject("books", books);
24         mv.setViewName("books");
25         return mv;
26     }
27 }
```

代码解释：

- 创建 Book 实体类，承载返回数据。
- 在 BookController 中，第 11~21 行构建返回数据，第 22~25 行创建返回 ModelAndView，设置视图名为 books，返回数据为所创建的 List 集合。

4. 创建视图

在 resources 目录下的 templates 目录中创建 books.html，具体代码如下：

```
1  <!DOCTYPE html>
2  <html lang="en" xmlns:th="http://www.thymeleaf.org">
3  <head>
4  <meta charset="UTF-8">
5  <title>图书列表</title>
6  </head>
7  <body>
8  <table border="1">
9  <tr>
10 <td>图书编号</td>
11 <td>图书名称</td>
12 <td>图书作者</td>
13 </tr>
14 <tr th:each="book:${books}">
15 <td th:text="${book.id}"></td>
16 <td th:text="${book.name}"></td>
17 <td th:text="${book.author}"></td>
18 </tr>
```

```
19  </table>
20  </body>
21  </html>
```

代码解释：

- 首先在第 2 行导入 Thymeleaf 的名称空间。
- 第 14~18 行通过遍历将 books 中的数据展示出来，Thymeleaf 中通过 th:each 进行集合遍历，通过 th:text 展示数据。

5. 运行

在浏览器地址栏中输入"http://localhost:8080/books"，即可看到运行结果，如图 3-1 所示。

图 3-1

本节重点介绍 Spring Boot 整合 Thymeleaf，并非 Thymeleaf 的基础用法，关于 Thymeleaf 的更多资料，可以查看 https://www.thymeleaf.org。

3.2 整合 FreeMarker

FreeMarker 是一个非常古老的模板引擎，可以用在 Web 环境或者非 Web 环境中。与 Thymeleaf 不同，FreeMarker 需要经过解析才能够在浏览器中展示出来。FreeMarker 不仅可以用来配置 HTML 页面模板，也可以作为电子邮件模板、配置文件模板以及源码模板等。Spring Boot 中对 FreeMarker 整合也提供了很好的支持，主要整合步骤如下：

1. 创建项目，添加依赖

首先创建 Spring Boot 项目，然后添加 spring-boot-starter-web 和 spring-boot-starter-freemarker 依赖，代码如下：

```
1  <dependency>
2      <groupId>org.springframework.boot</groupId>
3      <artifactId>spring-boot-starter-freemarker</artifactId>
4  </dependency>
5  <dependency>
6      <groupId>org.springframework.boot</groupId>
7      <artifactId>spring-boot-starter-web</artifactId>
8  </dependency>
```

2. 配置 FreeMarker

Spring Boot 对 FreeMarker 也提供了自动化配置类 FreeMarkerAutoConfiguration，相关的配置属性在 FreeMarkerProperties 中，FreeMarkerProperties 部分源码如下：

```
@ConfigurationProperties(prefix = "spring.freemarker")
public class FreeMarkerProperties extends AbstractTemplateViewResolverProperties {
    public static final String DEFAULT_TEMPLATE_LOADER_PATH = "classpath:/templates/";
    public static final String DEFAULT_PREFIX = "";
    public static final String DEFAULT_SUFFIX = ".ftl";
    …
    …
}
```

从该默认配置中可以看到，FreeMarker 默认模板位置和 Thymeleaf 一样，都在 classpath:/templates/中，默认文件后缀是.ftl。开发者可以在 application.properties 中对这些默认配置进行修改，部分常见配置如下：

```
#HttpServletRequest 的属性是否可以覆盖 controller 中 model 的同名项
spring.freemarker.allow-request-override=false
#HttpSession 的属性是否可以覆盖 controller 中 model 的同名项
spring.freemarker.allow-session-override=false
#是否开启缓存
spring.freemarker.cache=false
#模板文件编码
spring.freemarker.charset=UTF-8
#是否检查模板位置
spring.freemarker.check-template-location=true
#Content-Type 的值
spring.freemarker.content-type=text/html
#是否将 HttpServletRequest 中的属性添加到 Model 中
spring.freemarker.expose-request-attributes=false
#是否将 HttpSession 中的属性添加到 Model 中
spring.freemarker.expose-session-attributes=false
#模板文件后缀
spring.freemarker.suffix=.ftl
#模板文件位置
spring.freemarker.template-loader-path=classpath:/templates/
```

3. 配置控制器

控制器和 Thymeleaf 中的控制器一样，这里不再重复。

4. 创建视图

按照配置文件，在 resources/templates 目录下创建 books.ftl 文件，内容如下：

```
<!DOCTYPE html>
<html lang="en">
<head>
<meta charset="UTF-8">
<title>图书列表</title>
```

```
6    </head>
7    <body>
8    <table border="1">
9    <tr>
10   <td>图书编号</td>
11   <td>图书名称</td>
12   <td>图书作者</td>
13   </tr>
14   <#if books ??&&(books?size>0)>
15   <#list books as book>
16   <tr>
17   <td>${book.id}</td>
18   <td>${book.name}</td>
19   <td>${book.author}</td>
20   </tr>
21   </#list>
22   </#if>
23   </table>
24   </body>
25   </html>
```

代码解释：

- 第 14 行首先判断 model 中的 books 不为空并且 books 中有数据，然后进行遍历。
- 第 15~21 行表示遍历 books 集合，将集合中的数据通过表格展示出来。

5. 运行

在浏览器中输入 "http://localhost:8080/books"，即可看到运行结果，如图 3-2 所示。

本节重点介绍 Spring Boot 整合 FreeMarker，并非 FreeMarker 的基础用法，关于 FreeMarker 的更多资料，可以查看 https://freemarker.apache.org/ 。

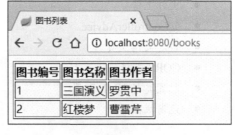

图 3-2

3.3 小　　结

本章向读者介绍了 Spring Boot 整合视图层技术，选择了两个具有代表性的例子：Thymeleaf 和 FreeMarker。开发者用到其他模板技术时，整合方式和 Thymeleaf、FreeMarker 基本一致。如果开发者使用的是目前流行的前后端分离技术，那么在开发过程中不需要整合视图层技术，后端直接提供接口即可。第 4 章将向读者介绍 Spring Boot 整合 Web 开发的其他细节。

第 4 章

Spring Boot 整合 Web 开发

本章概要

- 返回 JSON 数据
- 静态资源访问
- 文件上传
- @ControllerAdvice
- 自定义错误页
- CORS 支持
- 配置类与 XML 配置
- 注册拦截器
- 启动系统任务
- 整合 Servlet、Filter 和 Listener
- 路径映射
- 配置 AOP
- 自定义欢迎页
- 自定义 favicon
- 除去某个自动配置

4.1 返回 JSON 数据

4.1.1 默认实现

JSON 是目前主流的前后端数据传输方式，Spring MVC 中使用消息转换器 HttpMessageConverter 对 JSON 的转换提供了很好的支持，在 Spring Boot 中更进一步，对相关配置做了更进一步的简化。默认情况下，当开发者新创建一个 Spring Boot 项目后，添加 Web 依赖，代码如下：

```xml
<dependency>
<groupId>org.springframework.boot</groupId>
<artifactId>spring-boot-starter-web</artifactId>
</dependency>
```

这个依赖中默认加入了 jackson-databind 作为 JSON 处理器，此时不需要添加额外的 JSON 处理器就能返回一段 JSON 了。创建一个 Book 实体类：

```java
public class Book {
    private String name;
    private String author;
    @JsonIgnore
    private Float price;
    @JsonFormat(pattern = "yyyy-MM-dd")
    private Date publicationDate;
    //省略getter/setter
}
```

然后创建 BookController，返回 Book 对象即可：

```java
@Controller
public class BookController {
    @GetMapping("/book")
    @ResponseBody
    public Book book() {
        Book book = new Book();
        book.setAuthor("罗贯中");
        book.setName("三国演义");
        book.setPrice(30f);
        book.setPublicationDate(new Date());
        return book;
    }
}
```

当然，如果需要频繁地用到@ResponseBody 注解，那么可以采用@RestController 组合注解代替@Controller 和@ResponseBody，代码如下：

```java
@RestController
public class BookController {
    @GetMapping("/book")
    public Book book() {
        Book book = new Book();
        book.setAuthor("罗贯中");
        book.setName("三国演义");
        book.setPrice(30f);
        book.setPublicationDate(new Date());
        return book;
    }
}
```

此时，在浏览器中输入"http://localhost:8080/book"，即可看到返回了 JSON 数据，如图 4-1 所示。

图 4-1

这是 Spring Boot 自带的处理方式。如果采用这种方式，那么对于字段忽略、日期格式化等常见需求都可以通过注解来解决。

这是通过 Spring 中默认提供的 MappingJackson2HttpMessageConverter 来实现的，当然开发者在这里也可以根据实际需求自定义 JSON 转换器。

4.1.2 自定义转换器

常见的 JSON 处理器除了 jackson-databind 之外，还有 Gson 和 fastjson，这里针对常见用法分别举例。

1. 使用 Gson

Gson 是 Google 的一个开源 JSON 解析框架。使用 Gson，需要先除去默认的 jackson-databind，然后加入 Gson 依赖，代码如下：

```
<dependency>
    <groupId>org.springframework.boot</groupId>
    <artifactId>spring-boot-starter-web</artifactId>
    <exclusions>
        <exclusion>
            <groupId>com.fasterxml.jackson.core</groupId>
            <artifactId>jackson-databind</artifactId>
        </exclusion>
    </exclusions>
</dependency>
<dependency>
    <groupId>com.google.code.gson</groupId>
    <artifactId>gson</artifactId>
</dependency>
```

由于 Spring Boot 中默认提供了 Gson 的自动转换类 GsonHttpMessageConvertersConfiguration，因此 Gson 的依赖添加成功后，可以像使用 jackson-databind 那样直接使用 Gson。但是在 Gson 进行转换时，如果想对日期数据进行格式化，那么还需要开发者自定义 HttpMessageConverter。自定义 HttpMessageConverter 可以通过如下方式。

首先看 GsonHttpMessageConvertersConfiguration 中的一段源码：

```
@Bean
@ConditionalOnMissingBean
public GsonHttpMessageConverter gsonHttpMessageConverter(Gson gson) {
    GsonHttpMessageConverter converter = new GsonHttpMessageConverter();
```

```
5      converter.setGson(gson);
6      return converter;
7  }
```

@ConditionalOnMissingBean 注解表示当项目中没有提供 GsonHttpMessageConverter 时才会使用默认的 GsonHttpMessageConverter，所以开发者只需要提供一个 GsonHttpMessageConverter 即可，代码如下：

```
1  @Configuration
2  public class GsonConfig {
3      @Bean
4      GsonHttpMessageConverter gsonHttpMessageConverter() {
5          GsonHttpMessageConverter converter = new GsonHttpMessageConverter();
6          GsonBuilder builder = new GsonBuilder();
7          builder.setDateFormat("yyyy-MM-dd");
8          builder.excludeFieldsWithModifiers(Modifier.PROTECTED);
9          Gson gson = builder.create();
10         converter.setGson(gson);
11         return converter;
12     }
13 }
```

代码解释：

- 开发者自己提供一个 GsonHttpMessageConverter 的实例。
- 设置 Gson 解析时日期的格式。
- 设置 Gson 解析时修饰符为 protected 的字段被过滤掉。
- 创建 Gson 对象放入 GsonHttpMessageConverter 的实例中并返回 converter。

此时，将 Book 类中的 price 字段的修饰符改为 protected，代码如下：

```
1  public class Book {
2      private String name;
3      private String author;
4      protected Float price;
5      private Date publicationDate;
6      //省略 getter/setter
7  }
```

最后，在浏览器中输入"http://localhost:8080/book"，即可看到运行结果，如图 4-2 所示。

图 4-2

2. 使用 fastjson

fastjson 是阿里巴巴的一个开源 JSON 解析框架，是目前 JSON 解析速度最快的开源框架，该框架也可以集成到 Spring Boot 中。不同于 Gson，fastjson 继承完成之后并不能立马使用，需要开发者提供相应的 HttpMessageConverter 后才能使用，集成 fastjson 的步骤如下。

首先除去 jackson-databind 依赖，引入 fastjson 依赖：

```
<dependency>
<groupId>org.springframework.boot</groupId>
<artifactId>spring-boot-starter-web</artifactId>
<exclusions>
<exclusion>
<groupId>com.fasterxml.jackson.core</groupId>
<artifactId>jackson-databind</artifactId>
</exclusion>
</exclusions>
</dependency>
<dependency>
<groupId>com.alibaba</groupId>
<artifactId>fastjson</artifactId>
<version>1.2.47</version>
</dependency>
```

然后配置 fastjson 的 HttpMessageConverter：

```
@Configuration
public class MyFastJsonConfig {
    @Bean
    FastJsonHttpMessageConverter fastJsonHttpMessageConverter() {
        FastJsonHttpMessageConverter converter = new FastJsonHttpMessageConverter();
        FastJsonConfig config = new FastJsonConfig();
        config.setDateFormat("yyyy-MM-dd");
        config.setCharset(Charset.forName("UTF-8"));
        config.setSerializerFeatures(
                SerializerFeature.WriteClassName,
                SerializerFeature.WriteMapNullValue,
                SerializerFeature.PrettyFormat,
                SerializerFeature.WriteNullListAsEmpty,
                SerializerFeature.WriteNullStringAsEmpty
        );
        converter.setFastJsonConfig(config);
        return converter;
    }
}
```

代码解释：

- 自定义 MyFastJsonConfig，完成对 FastJsonHttpMessageConverter Bean 的提供。
- 第 7~15 行分别配置了 JSON 解析过程的一些细节，例如日期格式、数据编码、是否在生成的 JSON 中输出类名、是否输出 value 为 null 的数据、生成的 JSON 格式化、空集合输出[]而非 null、空字符串输出""而非 null 等基本配置。

MyFastJsonConfig 配置完成后，还需要配置一下响应编码，否则返回的 JSON 中文会乱码，在

application.properties 中添加如下配置：

```
spring.http.encoding.force-response=true
```

接下来提供 BookController 进行测试。BookController 和上一小节一致，运行成功后，在浏览器中输入"http://localhost:8080/book"，即可看到运行结果，如图 4-3 所示。

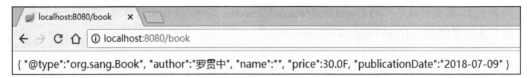

图 4-3

对于 FastJsonHttpMessageConverter 的配置，除了上面这种方式之外，还有另一种方式。

在 Spring Boot 项目中，当开发者引入 spring-boot-starter-web 依赖之后，该依赖又依赖了 spring-boot-autoconfigure，在这个自动化配置中，有一个 WebMvcAutoConfiguration 类提供了对 Spring MVC 最基本的配置，如果某一项自动化配置不满足开发需求，开发者可以针对该项自定义配置，只需要实现 WebMvcConfigurer 接口即可（在 Spring 5.0 之前是通过继承 WebMvcConfigurerAdapter 类来实现的），代码如下：

```
@Configuration
public class MyWebMvcConfig implements WebMvcConfigurer {
    @Override
    public void configureMessageConverters(List<HttpMessageConverter<?>> converters) {
        FastJsonHttpMessageConverter converter = new FastJsonHttpMessageConverter();
        FastJsonConfig config = new FastJsonConfig();
        config.setDateFormat("yyyy-MM-dd");
        config.setCharset(Charset.forName("UTF-8"));
        config.setSerializerFeatures(
            SerializerFeature.WriteClassName,
            SerializerFeature.WriteMapNullValue,
            SerializerFeature.PrettyFormat,
            SerializerFeature.WriteNullListAsEmpty,
            SerializerFeature.WriteNullStringAsEmpty
        );
        converter.setFastJsonConfig(config);
        converters.add(converter);
    }
}
```

代码解释：

- 自定义 MyWebMvcConfig 类并实现 WebMvcConfigurer 接口中的 configureMessageConverters 方法。
- 将自定义的 FastJsonHttpMessageConverter 加入 converters 中。

> **注　意**
>
> 如果使用了 Gson，也可以采用这种方式配置，但是不推荐。因为当项目中没有 GsonHttpMessageConverter 时，Spring Boot 自己会提供一个 GsonHttpMessageConverter，此时重写 configureMessageConverters 方法，参数 converters 中已经有 GsonHttpMessageConverter 的实例了，需要替换已有的 GsonHttpMessageConverter 实例，操作比较麻烦，所以对于 Gson，推荐直接提供 GsonHttpMessageConverter。

4.2　静态资源访问

在 Spring MVC 中，对于静态资源都需要开发者手动配置静态资源过滤。Spring Boot 中对此也提供了自动化配置，可以简化静态资源过滤配置。

4.2.1　默认策略

Spring Boot 中对于 Spring MVC 的自动化配置都在 WebMvcAutoConfiguration 类中，因此对于默认的静态资源过滤策略可以从这个类中一窥究竟。

在 WebMvcAutoConfiguration 类中有一个静态内部类 WebMvcAutoConfigurationAdapter，实现了 4.1 节提到的 WebMvcConfigurer 接口。WebMvcConfigurer 接口中有一个方法 addResourceHandlers，是用来配置静态资源过滤的。方法在 WebMvcAutoConfigurationAdapter 类中得到了实现，部分核心代码如下：

```
 1  public void addResourceHandlers(ResourceHandlerRegistry registry) {
 2    …
 3    …
 4    String staticPathPattern = this.mvcProperties.getStaticPathPattern();
 5      if (!registry.hasMappingForPattern(staticPathPattern)) {
 6        customizeResourceHandlerRegistration(
 7            registry.addResourceHandler(staticPathPattern)
 8              .addResourceLocations(getResourceLocations(
 9                  this.resourceProperties.getStaticLocations()))
10              .setCachePeriod(getSeconds(cachePeriod))
11              .setCacheControl(cacheControl));
12      }
13  }
```

Spring Boot 在这里进行了默认的静态资源过滤配置，其中 staticPathPattern 默认定义在 WebMvcProperties 中，定义内容如下：

```
 1  private String staticPathPattern = "/**";
```

this.resourceProperties.getStaticLocations()获取到的默认静态资源位置定义在 ResourceProperties

中，代码如下：

```
private static final String[] CLASSPATH_RESOURCE_LOCATIONS = {
"classpath:/META-INF/resources/", "classpath:/resources/",
"classpath:/static/", "classpath:/public/" };
```

在 getResourceLocations 方法中，对这 4 个静态资源位置做了扩充，代码如下：

```
static String[] getResourceLocations(String[] staticLocations) {
String[] locations = new String[staticLocations.length+ SERVLET_LOCATIONS.length];
System.arraycopy(staticLocations, 0, locations, 0, staticLocations.length);
System.arraycopy(SERVLET_LOCATIONS, 0, locations, staticLocations.length,
SERVLET_LOCATIONS.length);
    return locations;
}
```

其中，SERVLET_LOCATIONS 的定义是一个 { "/" }。

综上可以看到，Spring Boot 默认会过滤所有的静态资源，而静态资源的位置一共有 5 个，分别是"classpath:/META-INF/resources/"、"classpath:/resources/"、"classpath:/static/"、"classpath:/public/"以及"/"，也就是说，开发者可以将静态资源放到这 5 个位置中的任意一个。注意，按照定义的顺序，5 个静态资源位置的优先级依次降低。但是一般情况下，Spring Boot 项目不需要 webapp 目录，所以第 5 个"/"可以暂不考虑。

在一个新创建的 Spring Boot 项目中，添加了 spring-boot-starter-web 依赖之后，在 resources 目录下分别创建 4 个目录，4 个目录中放入同名的静态资源（如图 4-4 所示，数字表示不同位置资源的优先级）。

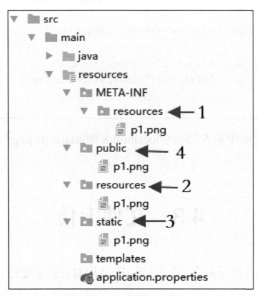

图 4-4

此时，在浏览器中输入 "http://localhost:8080/p1.png" 即可看到 classpath:/META-INF/resources/目录下的 p1.png，如果将 classpath:/META-INF/resources/目录下的 p1.png 删除，就会访问到 classpath:/resources/目录下的 p1.png，以此类推。

如果开发者使用 IntelliJ IDEA 创建 Spring Boot 项目，就会默认创建出 classpath:/static/ 目录，静态资源一般放在这个目录下即可。

4.2.2 自定义策略

如果默认的静态资源过滤策略不能满足开发需求，也可以自定义静态资源过滤策略，自定义静态资源过滤策略有以下两种方式：

1. 在配置文件中定义

可以在 application.properties 中直接定义过滤规则和静态资源位置，代码如下：

```
1  spring.mvc.static-path-pattern=/static/**
2  spring.resources.static-locations=classpath:/static/
```

过滤规则为/static/**，静态资源位置为 classpath:/static/。

重新启动项目，在浏览器中输入"http://localhost:8080/static/p1.png"，即可看到 classpath:/static/ 目录下的资源。

2. Java 编码定义

也可以通过 Java 编码方式来定义，此时只需要实现 WebMvcConfigurer 接口即可，然后实现该接口的 addResourceHandlers 方法，代码如下：

```
1  @Configuration
2  public class MyWebMvcConfig implements WebMvcConfigurer {
3      @Override
4      public void addResourceHandlers(ResourceHandlerRegistry registry) {
5          registry
6                  .addResourceHandler("/static/**")
7                  .addResourceLocations("classpath:/static/");
8      }
9  }
```

重新启动项目，在浏览器中输入"http://localhost:8080/static/p1.png"，即可看到 classpath:/static/ 目录下的资源。

4.3 文件上传

Spring MVC 对文件上传做了简化，在 Spring Boot 中对此做了更进一步的简化，文件上传更为方便。

Java 中的文件上传一共涉及两个组件，一个是 CommonsMultipartResolver，另一个是 StandardServletMultipartResolver，其中 CommonsMultipartResolver 使用 commons-fileupload 来处理 multipart 请求，而 StandardServletMultipartResolver 则是基于 Servlet 3.0 来处理 multipart 请求的，因此若使用 StandardServletMultipartResolver，则不需要添加额外的 jar 包。Tomcat 7.0 开始就支持

Servlet 3.0 了，而 Spring Boot 2.0.4 内嵌的 Tomcat 为 Tomcat 8.5.32，因此可以直接使用 StandardServletMultipartResolver。而在 Spring Boot 提供的文件上传自动化配置类 MultipartAutoConfiguration 中，默认也是采用 StandardServletMultipartResolver，部分源码如下：

```
public class MultipartAutoConfiguration {
  …
  …
   @Bean(name = DispatcherServlet.MULTIPART_RESOLVER_BEAN_NAME)
   @ConditionalOnMissingBean(MultipartResolver.class)
   public StandardServletMultipartResolver multipartResolver() {
      StandardServletMultipartResolver multipartResolver = new StandardServletMultipartResolver();
multipartResolver
.setResolveLazily(this.multipartProperties.isResolveLazily());
      return multipartResolver;
   }
}
```

根据这里的配置可以看出，如果开发者没有提供 MultipartResolver，那么默认采用的 MultipartResolver 就是 StandardServletMultipartResolver。因此，在 Spring Boot 中上传文件甚至可以做到零配置。下面来看具体上传过程。

4.3.1 单文件上传

首先创建一个 Spring Boot 项目并添加 spring-boot-starter-web 依赖。

然后在 resources 目录下的 static 目录中创建一个 upload.html 文件，内容如下：

```
<!DOCTYPE html>
<html lang="en">
<head>
<meta charset="UTF-8">
<title>Title</title>
</head>
<body>
<form action="/upload" method="post" enctype="multipart/form-data">
<input type="file" name="uploadFile" value="请选择文件">
<input type="submit" value="上传">
</form>
</body>
</html>
```

这是一个很简单的文件上传页面，上传接口是/upload，注意请求方法是 post，enctype 是 multipart/form-data。

接着创建文件上传处理接口，代码如下：

```
@RestController
public class FileUploadController {
   SimpleDateFormat sdf = new SimpleDateFormat("yyyy/MM/dd/");
```

```
4       @PostMapping("/upload")
5       public String upload(MultipartFile uploadFile, HttpServletRequest req) {
6           String realPath =
7   req.getSession().getServletContext().getRealPath("/uploadFile/");
8           String format = sdf.format(new Date());
9           File folder = new File(realPath + format);
10          if (!folder.isDirectory()) {
11              folder.mkdirs();
12          }
13          String oldName = uploadFile.getOriginalFilename();
14          String newName = UUID.randomUUID().toString() +
15   oldName.substring(oldName.lastIndexOf("."), oldName.length());
16          try {
17              uploadFile.transferTo(new File(folder, newName));
18              String filePath = req.getScheme() + "://" + req.getServerName() + ":" +
19   req.getServerPort() + "/uploadFile/" + format + newName;
20              return filePath;
21          } catch (IOException e) {
22              e.printStackTrace();
23          }
24          return "上传失败!";
25      }
26  }
```

代码解释：

- 第 6~12 代码表示规划上传文件的保存路径为项目运行目录下的 uploadFile 文件夹，并在文件夹中通过日期对所上传的文件归类保存。
- 第 13~15 行代码表示给上传的文件重命名，这是为了避免文件重名。
- 第 17 行是文件保存操作。
- 第 18~20 行是生成上传文件的访问路径，并将访问路径返回。

最后在浏览器中进行测试。

运行项目，在浏览器中输入"http://localhost:8080/upload.html"进行文件上传，如图 4-5 所示。

图 4-5

单击"请选择文件"按钮上传文件，文件上传成功后，会返回上传文件的访问路径，如图 4-6 所示。

图 4-6

用这个路径就可以看到刚刚上传的图片，如图 4-7 所示。

图 4-7

在 4.2 节中向读者介绍过静态资源位置除了 classpath 下面的 4 个路径之外，还有一个" /"，因此这里的图片虽然是静态资源却可以直接访问到。

至此，一个简单的图片上传逻辑就完成了，对于开发者而言，只需要专注于图片上传的业务逻辑，而不需要在配置上花费太多时间。

当然，如果开发者需要对图片上传的细节进行配置，也是允许的，代码如下：

```
1  spring.servlet.multipart.enabled=true
2  spring.servlet.multipart.file-size-threshold=0
3  spring.servlet.multipart.location=E:\\temp
4  spring.servlet.multipart.max-file-size=1MB
5  spring.servlet.multipart.max-request-size=10MB
6  spring.servlet.multipart.resolve-lazily=false
```

代码解释：

- 第 1 行表示是否开启文件上传支持，默认为 true。
- 第 2 行表示文件写入磁盘的阈值，默认为 0。
- 第 3 行表示上传文件的临时保存位置。
- 第 4 行表示上传的单个文件的最大大小，默认为 1MB。
- 第 5 行表示多文件上传时文件的总大小，默认为 10MB。
- 第 6 行表示文件是否延迟解析，默认为 false。

4.3.2　多文件上传

多文件上传和单文件上传基本一致，首先修改 HTML 文件，代码如下：

```html
1   <!DOCTYPE html>
2   <html lang="en">
3   <head>
4   <meta charset="UTF-8">
5   <title>Title</title>
6   </head>
7   <body>
8   <form action="/uploads" method="post" enctype="multipart/form-data">
9   <input type="file" name="uploadFiles" value="请选择文件" multiple>
10  <input type="submit" value="上传">
11  </form>
12  </body>
13  </html>
```

然后修改控制器，代码如下：

```java
1   @PostMapping("/uploads")
2   public String upload(MultipartFile[] uploadFiles, HttpServletRequest req) {
3       //遍历 uploadFiles 数组分别存储
4   }
```

控制器里边的核心逻辑和单文件上传是一样的，只是多一个遍历的步骤。

4.4 @ControllerAdvice

顾名思义，@ControllerAdvice 就是@Controller 的增强版。@ControllerAdvice 主要用来处理全局数据，一般搭配@ExceptionHandler、@ModelAttribute 以及@InitBinder 使用。

4.4.1 全局异常处理

@ControllerAdvice 最常见的使用场景就是全局异常处理。在 4.3 节向读者介绍过文件上传大小限制的配置，如果用户上传的文件超过了限制大小，就会抛出异常，此时可以通过@ControllerAdvice 结合@ExceptionHandler 定义全局异常捕获机制，代码如下：

```java
1   @ControllerAdvice
2   public class CustomExceptionHandler {
3       @ExceptionHandler(MaxUploadSizeExceededException.class)
4       public void uploadException(MaxUploadSizeExceededException e,
5   HttpServletResponse resp) throws IOException {
6           resp.setContentType("text/html;charset=utf-8");
7           PrintWriter out = resp.getWriter();
8           out.write("上传文件大小超出限制!");
9           out.flush();
10          out.close();
11      }
12  }
```

只需在系统中定义 CustomExceptionHandler 类，然后添加@ControllerAdvice 注解即可。当系统启动时，该类就会被扫描到 Spring 容器中，然后定义 uploadException 方法，在该方法上添加了 @ExceptionHandler 注解，其中定义的 MaxUploadSizeExceededException.class 表明该方法用来处理 MaxUploadSizeExceededException 类型的异常。如果想让该方法处理所有类型的异常，只需将 MaxUploadSizeExceededException 改为 Exception 即可。方法的参数可以有异常实例、HttpServletResponse 以及 HttpServletRequest、Model 等，返回值可以是一段 JSON、一个 ModelAndView、一个逻辑视图名等。此时，上传一个超大文件会有错误提示给用户，如图 4-8 所示。

图 4-8

如果返回参数是一个 ModelAndView，假设使用的页面模板为 Thymeleaf（注意添加 Thymeleaf 相关依赖），此时异常处理方法定义如下：

```
1   @ControllerAdvice
2   public class CustomExceptionHandler {
3       @ExceptionHandler(MaxUploadSizeExceededException.class)
4       public ModelAndView uploadException(MaxUploadSizeExceededException e) throws
5   IOException {
6           ModelAndView mv = new ModelAndView();
7           mv.addObject("msg", "上传文件大小超出限制!");
8           mv.setViewName("error");
9           return mv;
10      }
11  }
```

然后在 resources/templates 目录下创建 error.html 文件，内容如下：

```
1   <!DOCTYPE html>
2   <html lang="en" xmlns:th="http://www.thymeleaf.org">
3   <head>
4   <meta charset="UTF-8">
5   <title>Title</title>
6   </head>
7   <body>
8   <div th:text="${msg}"></div>
9   </body>
10  </html>
```

此时上传出错效果与图 4-8 一致。

4.4.2 添加全局数据

@ControllerAdvice 是一个全局数据处理组件，因此也可以在@ControllerAdvice 中配置全局数据，使用@ModelAttribute 注解进行配置，代码如下：

```
@ControllerAdvice
public class GlobalConfig {
    @ModelAttribute(value = "info")
    public Map<String,String> userInfo() {
        HashMap<String, String> map = new HashMap<>();
        map.put("username", "罗贯中");
        map.put("gender", "男");
        return map;
    }
}
```

代码解释：

- 在全局配置中添加 userInfo 方法，返回一个 map。该方法有一个注解@ModelAttribute，其中的 value 属性表示这条返回数据的 key，而方法的返回值是返回数据的 value。
- 此时在任意请求的 Controller 中，通过方法参数中的 Model 都可以获取 info 的数据。

Controller 示例代码如下：

```
@GetMapping("/hello")
@ResponseBody
public void hello(Model model) {
    Map<String, Object> map = model.asMap();
    Set<String> keySet = map.keySet();
    Iterator<String> iterator = keySet.iterator();
    while (iterator.hasNext()) {
        String key = iterator.next();
        Object value = map.get(key);
        System.out.println(key + ">>>>>" + value);
    }
}
```

在请求方法中，将 Model 中的数据打印出来，如图 4-9 所示。

```
2018-07-11 12:09:07.812  INFO 12604 ---
info>>>>>{gender=男, username=罗贯中}
```

图 4-9

4.4.3 请求参数预处理

@ControllerAdvice 结合@InitBinder 还能实现请求参数预处理，即将表单中的数据绑定到实体类上时进行一些额外处理。

例如有两个实体类 Book 和 Author,代码如下:

```
public class Book {
    private String name;
    private String author;
    //省略getter/setter
}
public class Author {
    private String name;
    private int age;
    //省略getter/setter
}
```

在 Controller 上需要接收两个实体类的数据,Controller 中的方法定义如下:

```
@GetMapping("/book")
@ResponseBody
public String book(Book book,Author author) {
    return book.toString() + ">>>" + author.toString();
}
```

此时在参数传递时,两个实体类中的 name 属性会混淆,@ControllerAdvice 结合@InitBinder 可以顺利解决该问题。配置步骤如下。

首先给 Controller 中方法的参数添加@ModelAttribute 注解,代码如下:

```
@GetMapping("/book")
@ResponseBody
public String book(@ModelAttribute("b") Book book, @ModelAttribute("a") Author author) {
    return book.toString() + ">>>" + author.toString();
}
```

然后配置@ControllerAdvice,代码如下:

```
@ControllerAdvice
public class GlobalConfig {
    @InitBinder("b")
    public void init(WebDataBinder binder) {
        binder.setFieldDefaultPrefix("b.");
    }
    @InitBinder("a")
    public void init2(WebDataBinder binder) {
        binder.setFieldDefaultPrefix("a.");
    }
}
```

代码解释:

- 在 GlobalConfig 类中创建两个方法,第一个 @InitBinder("b") 表示该方法是处理@ModelAttribute("b") 对应的参数的,第二个 @InitBinder("a") 表示该方法是处理@ModelAttribute("a")对应的参数的。

- 在每个方法中给相应的 Field 设置一个前缀，然后在浏览器中请求 http://localhost:8080/book?b.name=三国演义&b.author=罗贯中&a.name=曹雪芹&a.age=48，即可成功地区分出 name 属性。
- 在 WebDataBinder 对象中，还可以设置允许的字段、禁止的字段、必填字段以及验证器等。

4.5 自定义错误页

4.4 节向读者介绍了 Spring Boot 中的全局异常处理。在处理异常时，开发者可以根据实际情况返回不同的页面，但是这种异常处理方式一般用来处理应用级别的异常，有一些容器级别的错误就处理不了，例如 Filter 中抛出异常，使用@ControllerAdvice 定义的全局异常处理机制就无法处理。因此，Spring Boot 中对于异常的处理还有另外的方式，这就是本节要介绍的内容。

在 Spring Boot 中，默认情况下，如果用户在发起请求时发生了 404 错误，Spring Boot 会有一个默认的页面展示给用户，如图 4-10 所示。

图 4-10

如果发起请求时发生了 500 错误，Spring Boot 也会有一个默认的页面展示给用户，如图 4-11 所示。

图 4-11

事实上，Spring Boot 在返回错误信息时不一定返回 HTML 页面，而是根据实际情况返回 HTML 页面或者一段 JSON（若开发者发起 Ajax 请求，则错误信息是一段 JSON）。对于开发者而言，这一段 HTML 或者 JSON 都能够自由定制。

Spring Boot 中的错误默认是由 BasicErrorController 类来处理的，该类中的核心方法主要有两个：

```
1   @RequestMapping(produces = "text/html")
2   public ModelAndView errorHtml(HttpServletRequest request,
3       HttpServletResponse response) {
4     HttpStatus status = getStatus(request);
5     Map<String, Object> model = Collections.unmodifiableMap(getErrorAttributes(
6         request, isIncludeStackTrace(request, MediaType.TEXT_HTML)));
7     response.setStatus(status.value());
8     ModelAndView modelAndView = resolveErrorView(request, response, status, model);
9     return (modelAndView != null ? modelAndView : new ModelAndView("error", model));
10  }
11  @RequestMapping
12  @ResponseBody
13  public ResponseEntity<Map<String, Object>> error(HttpServletRequest request) {
14    Map<String, Object> body = getErrorAttributes(request,
15        isIncludeStackTrace(request, MediaType.ALL));
16    HttpStatus status = getStatus(request);
17    return new ResponseEntity<>(body, status);
18  }
```

其中，errorHtml 方法用来返回错误 HTML 页面，error 用来返回错误 JSON，具体返回的是 HTML 还是 JSON，则要看请求头的 Accept 参数。返回 JSON 的逻辑很简单，不必过多介绍，返回 HTML 的逻辑稍微有些复杂，在 errorHtml 方法中，通过调用 resolveErrorView 方法来获取一个错误视图的 ModelAndView。而 resolveErrorView 方法的调用最终会来到 DefaultErrorViewResolver 类中。

DefaultErrorViewResolver 类是 Spring Boot 中默认的错误信息视图解析器，部分源码如下：

```
1   public class DefaultErrorViewResolver implements ErrorViewResolver, Ordered {
2     private static final Map<Series, String> SERIES_VIEWS;
3     static {
4       Map<Series, String> views = new EnumMap<>(Series.class);
5       views.put(Series.CLIENT_ERROR, "4xx");
6       views.put(Series.SERVER_ERROR, "5xx");
7       SERIES_VIEWS = Collections.unmodifiableMap(views);
8     }
9     …
10    …
11    private ModelAndView resolve(String viewName, Map<String, Object> model) {
12      String errorViewName = "error/" + viewName;
13      TemplateAvailabilityProvider provider = this.templateAvailabilityProviders
14          .getProvider(errorViewName, this.applicationContext);
15      if (provider != null) {
16        return new ModelAndView(errorViewName, model);
17      }
18      return resolveResource(errorViewName, model);
19  }
20  …
21  …
22  }
```

从这一段源码中可以看到，Spring Boot 默认是在 error 目录下查找 4xx、5xx 的文件作为错误视图，当找不到时会回到 errorHtml 方法中，然后使用 error 作为默认的错误页面视图名，如果名为

error 的视图也找不到，用户就会看到本节一开始展示的两个错误提示页面。整个错误处理流程大致就是这样的。

4.5.1 简单配置

通过上面的介绍，读者可能已经发现，要自定义错误页面其实很简单，提供 4xx 和 5xx 页面即可。如果开发者不需要向用户展示详细的错误信息，那么可以把错误信息定义成静态页面，直接在 resources/static 目录下创建 error 目录，然后在 error 目录中创建错误展示页面。错误展示页面的命名规则有两种：一种是 4xx.html、5xx.html；另一种是直接使用响应码命名文件，例如 404.html、405.html、500.html。第二种命名方式划分得更细，当出错时，不同的错误会展示不同的错误页面，如图 4-12 所示。

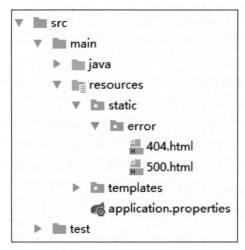

图 4-12

此时，当用户访问一个不存在的路径时，就会展示 404.html 页面中的内容，如图 4-13 所示。

图 4-13

修改 Controller，提供一个会抛异常的请求，代码如下：

```
@RestController
public class HelloController {
    @GetMapping("/hello")
    public String hello() {
        int i = 1 / 0;
        return "hello";
```

```
7       }
8   }
```

访问该接口，就会展示 500.html 中的内容，如图 4-14 所示。

图 4-14

这种定义都是使用了静态 HTML 页面，无法向用户展示完整的错误信息，若采用视图模板技术，则可以向用户展示更多的错误信息。如果要使用 HTML 模板，那么先引入模板相关的依赖，这里以 Thymeleaf 为例，Thymeleaf 页面模板默认处于 classpath:/templates/ 目录下，因此在该目录下先创建 error 目录，再创建错误展示页，如图 4-15 所示。

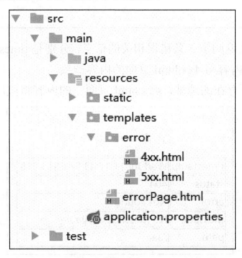

图 4-15

由于模板页面展示信息比较灵活，因此可以直接创建 4xx.html、5xx.html。以 4xx.html 页面为例，其内容如下：

```
1   <!DOCTYPE html>
2   <html lang="en" xmlns:th="http://www.thymeleaf.org/">
3   <head>
4   <meta charset="UTF-8">
5   <title>Title</title>
6   </head>
7   <body>
8   <table border="1">
9   <tr>
10  <td>timestamp</td>
11  <td th:text="${timestamp}"></td>
```

```
12  </tr>
13  <tr>
14  <td>status</td>
15  <td th:text="${status}"></td>
16  </tr>
17  <tr>
18  <td>error</td>
19  <td th:text="${error}"></td>
20  </tr>
21  <tr>
22  <td>message</td>
23  <td th:text="${message}"></td>
24  </tr>
25  <tr>
26  <td>path</td>
27  <td th:text="${path}"></td>
28  </tr>
29  </table>
30  </body>
31  </html>
```

Spring Boot 在这里一共返回了 5 条错误相关的信息，分别是 timestamp、status、error、message 以及 path。5xx.html 页面的内容与 4xx.html 页面的内容一致。

此时，用户访问一个不存在的地址，4xx.html 页面中的内容将被展示出来，如图 4-16 所示。

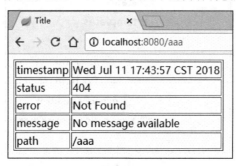

图 4-16

若用户访问一个会抛异常的地址，例如上文的/hello 接口，则会展示 5xx.html 页面的内容，如图 4-17 所示。

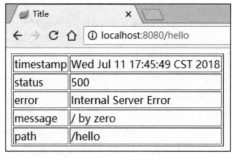

图 4-17

> **注 意**
>
> 若用户定义了多个错误页面，则响应码.html 页面的优先级高于 4xx.html、5xx.html 页面的优先级，即若当前是一个 404 错误，则优先展示 404.html 而不是 4xx.html；动态页面的优先级高于静态页面，即若 resources/templates 和 resources/static 下同时定义了 4xx.html，则优先展示 resources/templates/4xx.html。

4.5.2 复杂配置

上面这种配置还是不够灵活，只能定义 HTML 页面，无法处理 JSON 的定制。Spring Boot 中支持对 Error 信息的深度定制，接下来将从三个方面介绍深度定制：自定义 Error 数据、自定义 Error 视图以及完全自定义。

1. 自定义 Error 数据

自定义 Error 数据就是对返回的数据进行自定义。经过 4.5.1 小节的介绍，读者已经了解到 Spring Boot 返回的 Error 信息一共有 5 条，分别是 timestamp、status、error、message 以及 path。在 BasicErrorController 的 errorHtml 方法和 error 方法中，都是通过 getErrorAttributes 方法获取 Error 信息的。该方法最终会调用到 DefaultErrorAttributes 类的 getErrorAttributes 方法，而 DefaultErrorAttributes 类是在 ErrorMvcAutoConfiguration 中默认提供的。ErrorMvcAutoConfiguration 类的 errorAttributes 方法源码如下：

```
@Bean
@ConditionalOnMissingBean(value = ErrorAttributes.class, search =
SearchStrategy.CURRENT)
public DefaultErrorAttributes errorAttributes() {
  return new DefaultErrorAttributes(
      this.serverProperties.getError().isIncludeException());
}
```

从这段源码中可以看出，当系统没有提供 ErrorAttributes 时才会采用 DefaultErrorAttributes。因此自定义错误提示时，只需要自己提供一个 ErrorAttributes 即可，而 DefaultErrorAttributes 是 ErrorAttributes 的子类，因此只需要继承 DefaultErrorAttributes 即可，代码如下：

```
@Component
public class MyErrorAttribute extends DefaultErrorAttributes{
    @Override
    public Map<String, Object> getErrorAttributes(WebRequest webRequest, boolean includeStackTrace) {
        Map<String, Object> errorAttributes = super.getErrorAttributes(webRequest, includeStackTrace);
        errorAttributes.put("custommsg", "出错啦！");
        errorAttributes.remove("error");
        return errorAttributes;
    }
}
```

代码解释：

- 自定义 MyErrorAttribute 继承自 DefaultErrorAttributes，重写 DefaultErrorAttributes 中的 getErrorAttributes 方法。MyErrorAttribute 类添加@Component 注解，该类将被注册到 Spring 容器中。
- 第 6、7 行通过 super.getErrorAttributes 获取 Spring Boot 默认提供的错误信息，然后在此基础上添加 Error 信息或者移除 Error 信息。

此时，当系统抛出异常时，错误信息将被修改，以 4.5.1 节中的动态页面模板 404.html 为例，修改 404.html，代码如下：

```html
<!DOCTYPE html>
<html lang="en" xmlns:th="http://www.thymeleaf.org/">
<head>
<meta charset="UTF-8">
<title>Title</title>
</head>
<body>
<table border="1">
<tr>
<td>custommsg</td>
<td th:text="${custommsg}"></td>
</tr>
<tr>
<td>timestamp</td>
<td th:text="${timestamp}"></td>
</tr>
<tr>
<td>status</td>
<td th:text="${status}"></td>
</tr>
<tr>
<td>error</td>
<td th:text="${error}"></td>
</tr>
<tr>
<td>message</td>
<td th:text="${message}"></td>
</tr>
<tr>
<td>path</td>
<td th:text="${path}"></td>
</tr>
</table>
</body>
</html>
```

在第 9~12 行添加了 custommsg 属性，此时访问一个不存在的路径，就能看到自定义的 Error 信息，并且可以看到默认的 error 被移除了，如图 4-18 所示。

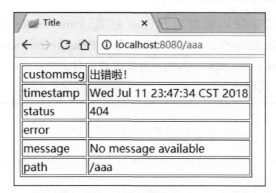

图 4-18

如果通过 Postman 等工具来发起这个请求,那么返回的 JSON 数据中也是如此,如图 4-19 所示。

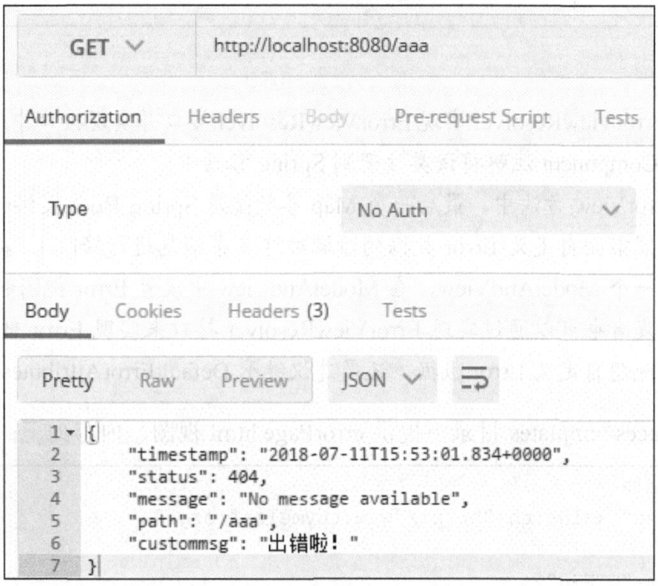

图 4-19

2. 自定义 Error 视图

Error 视图是展示给用户的页面,在 BasicErrorController 的 errorHtml 方法中调用 resolveErrorView 方法获取一个 ModelAndView 实例。resolveErrorView 方法是由 ErrorViewResolver 提供的,通过 ErrorMvcAutoConfiguration 类的源码可以看到 Spring Boot 默认采用的 ErrorViewResolver 是 DefaultErrorViewResolver。ErrorMvcAutoConfiguration 部分源码如下:

```
@Bean
@ConditionalOnBean(DispatcherServlet.class)
@ConditionalOnMissingBean
public DefaultErrorViewResolver conventionErrorViewResolver() {
  return new DefaultErrorViewResolver(this.applicationContext,
      this.resourceProperties);
}
```

从这一段源码可以看到，如果用户没有定义 ErrorViewResolver，那么默认使用的 ErrorViewResolver 是 DefaultErrorViewResolver，正是在 DefaultErrorViewResolver 中配置了默认去 error 目录下寻找 4xx.html、5xx.html。因此，开发者想要自定义 Error 视图，只需要提供自己的 ErrorViewResolver 即可，代码如下：

```java
@Component
public class MyErrorViewResolver implements ErrorViewResolver {
    @Override
    public ModelAndView resolveErrorView(HttpServletRequest request, HttpStatus status, Map<String, Object> model) {
        ModelAndView mv = new ModelAndView("errorPage");
        mv.addObject("custommsg", "出错啦!! ");
        mv.addAllObjects(model);
        return mv;
    }
}
```

代码解释：

- 自定义 MyErrorViewResolver 实现 ErrorViewResolver 接口并实现接口中的 resolveErrorView 方法，使用 @Component 注解将该类注册到 Spring 容器中。
- 在 resolveErrorView 方法中，最后一个 Map 参数就是 Spring Boot 提供的默认的 5 条 Error 信息（可以按照前面自定义 Error 数据的步骤对这 5 条消息进行修改）。在 resolveErrorView 方法中，返回一个 ModelAndView，在 ModelAndView 中设置 Error 视图和 Error 数据。
- 理论上，开发者也可以通过实现 ErrorViewResolver 接口来实现 Error 数据的自定义，但是如果只是单纯地想自定义 Error 数据，还是建议继承 DefaultErrorAttributes。

接下来在 resources/templates 目录下提供 errorPage.html 视图，内容如下：

```html
<!DOCTYPE html>
<html lang="en" xmlns:th="http://www.thymeleaf.org/">
<head>
<meta charset="UTF-8">
<title>Title</title>
</head>
<body>
<h3>errorPage</h3>
<table border="1">
<tr>
<td>custommsg</td>
<td th:text="${custommsg}"></td>
</tr>
<tr>
<td>timestamp</td>
<td th:text="${timestamp}"></td>
</tr>
<tr>
<td>status</td>
<td th:text="${status}"></td>
```

```
21  </tr>
22  <tr>
23  <td>error</td>
24  <td th:text="${error}"></td>
25  </tr>
26  <tr>
27  <td>message</td>
28  <td th:text="${message}"></td>
29  </tr>
30  <tr>
31  <td>path</td>
32  <td th:text="${path}"></td>
33  </tr>
34  </table>
35  </body>
36  </html>
```

在errorPage.html 中，除了展示 Spring Boot 提供的 5 条 Error 信息外，也展示了开发者自定义的 Error 信息。此时，无论请求发生 4xx 的错误（如图 4-20 所示）还是发生 5xx 的错误（如图 4-21 所示），都会来到 errorPage.html 页面。

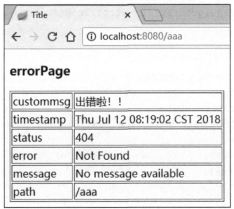

图 4-20

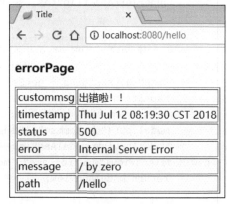

图 4-21

3. 完全自定义

前面提到的两种自定义方式都是对 BasicErrorController 类中的某个环节进行修补。查看 Error 自动化配置类 ErrorMvcAutoConfiguration，读者可以发现 BasicErrorController 本身只是一个默认的配置，相关源码如下：

```
1  public class ErrorMvcAutoConfiguration {
2  …
3      …
4      @Bean
5      @ConditionalOnMissingBean(value = ErrorController.class, search =
6      SearchStrategy.CURRENT)
7      public BasicErrorController basicErrorController(ErrorAttributes errorAttributes) {
8          return new BasicErrorController(errorAttributes,
```

```
9        this.serverProperties.getError(),
10       this.errorViewResolvers);
11   }
12   …
13   …
14 }
```

从这段源码中可以看到，若开发者没有提供自己的 ErrorController，则 Spring Boot 提供 BasicErrorController 作为默认的 ErrorController。因此，如果开发者需要更加灵活地对 Error 视图和数据进行处理，那么只需要提供自己的 ErrorController 即可。提供自己的 ErrorController 有两种方式：一种是实现 ErrorController 接口，另一种是直接继承 BasicErrorController。由于 ErrorController 接口只提供一个待实现的方法，而 BasicErrorController 已经实现了很多功能，因此这里选择第二种方式，即通过继承 BasicErrorController 来实现自己的 ErrorController。具体定义如下：

```
1  @Controller
2  public class MyErrorController extends BasicErrorController {
3      @Autowired
4      public MyErrorController(ErrorAttributes errorAttributes,
5                      ServerProperties serverProperties,
6                      List<ErrorViewResolver> errorViewResolvers) {
7          super(errorAttributes, serverProperties.getError(), errorViewResolvers);
8      }
9      @Override
10     public ModelAndView errorHtml(HttpServletRequest request,
11                     HttpServletResponse response) {
12         HttpStatus status = getStatus(request);
13         Map<String, Object> model = getErrorAttributes(
14                 request, isIncludeStackTrace(request, MediaType.TEXT_HTML));
15         model.put("custommsg", "出错啦！");
16         ModelAndView modelAndView =
17 new ModelAndView("myErrorPage", model, status);
18         return modelAndView;
19     }
20     @Override
21     public ResponseEntity<Map<String, Object>> error(HttpServletRequest request) {
22         Map<String, Object> body = getErrorAttributes(request,
23                 isIncludeStackTrace(request, MediaType.ALL));
24         body.put("custommsg", "出错啦！");
25         HttpStatus status = getStatus(request);
26         return new ResponseEntity<>(body, status);
27     }
28 }
```

代码解释：

- 自定义 MyErrorController 继承自 BasicErrorController 并添加 @Controller 注解，将 MyErrorController 注册到 Spring MVC 容器中。
- 由于 BasicErrorController 没有无参构造方法，因此在创建 BasicErrorController 实例时需要传递参数，在 MyErrorController 的构造方法上添加 @Autowired 注解注入所需参数。

- 参考 BasicErrorController 中的实现，重写 errorHtml 和 error 方法，对 Error 的视图和数据进行充分的自定义。

最后，在 resources/templates 目录下提供 myErrorPage.html 页面作为视图页面，代码如下：

```html
<!DOCTYPE html>
<html lang="en" xmlns:th="http://www.thymeleaf.org/">
<head>
<meta charset="UTF-8">
<title>Title</title>
</head>
<body>
<h3>myErrorPage</h3>
<table border="1">
<tr>
<td>custommsg</td>
<td th:text="${custommsg}"></td>
</tr>
<tr>
<td>timestamp</td>
<td th:text="${timestamp}"></td>
</tr>
<tr>
<td>status</td>
<td th:text="${status}"></td>
</tr>
<tr>
<td>error</td>
<td th:text="${error}"></td>
</tr>
<tr>
<td>message</td>
<td th:text="${message}"></td>
</tr>
<tr>
<td>path</td>
<td th:text="${path}"></td>
</tr>
</table>
</body>
</html>
```

访问一个不存在的页面，就可以看到自定义的错误提示了，如图 4-22 所示。

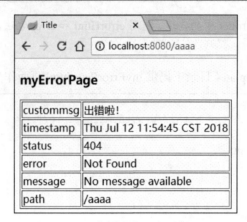

图 4-22

如果通过 Postman 等工具发起这个请求，那么返回数据为一段 JSON，如图 4-23 所示。

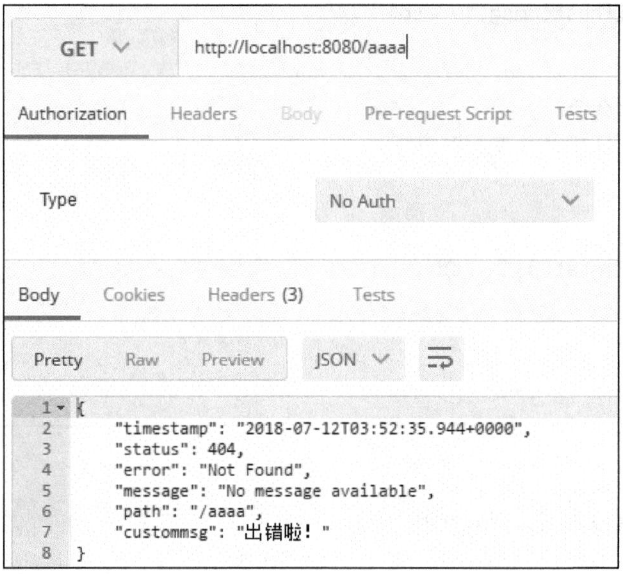

图 4-23

Spring Boot 中对异常的处理还是非常容易的，Spring Boot 虽然提供了非常丰富的自动化配置方案，但是也允许开发者根据实际情况进行完全的自定义，开发者在使用过程中可以结合具体情况选择合适的 Error 处理方案。

4.6 CORS 支持

CORS（Cross-Origin Resource Sharing）是由 W3C 制定的一种跨域资源共享技术标准，其目的就是为了解决前端的跨域请求。在 Java EE 开发中，最常见的前端跨域请求解决方案是 JSONP，但是 JSONP 只支持 GET 请求，这是一个很大的缺陷，而 CORS 则支持多种 HTTP 请求方法。以 CORS

中的 GET 请求为例，当浏览器发起请求时，请求头中携带了如下信息：

```
1  …
2  …
3  Host: localhost:8080
4  Origin: http://localhost:8081
5  Referer: http://localhost:8081/index.html
6  …
7  …
```

假如服务端支持 CORS，则服务端给出的响应信息如下：

```
1  …
2  …
3  Access-Control-Allow-Origin: http://localhost:8081
4  Content-Length: 20
5  Content-Type: text/plain;charset=UTF-8
6  Date: Thu, 12 Jul 2018 12:51:14 GMT
7  …
8  …
```

> **注 意**
>
> 响应头中有一个 Access-Control-Allow-Origin 字段，用来记录可以访问该资源的域。当浏览器收到这样的响应头信息之后，提取出 Access-Control-Allow-Origin 字段中的值，发现该值包含当前页面所在的域，就知道这个跨域是被允许的，因此就不再对前端的跨域请求进行限制。这就是 GET 请求的整个跨域流程，在这个过程中，前端请求的代码不需要修改，主要是后端进行处理。这个流程主要是针对 GET、POST 以及 HEAD 请求，并且没有自定义请求头，如果用户发起一个 DELETE 请求、PUT 请求或者自定义了请求头，流程就会稍微复杂一些。

以 DELETE 请求为例，当前端发起一个 DELETE 请求时，这个请求的处理会经过两个步骤。

第一步：发送一个 OPTIONS 请求。代码如下：

```
1  …
2  …
3  Access-Control-Request-Method DELETE
4  Connection keep-alive
5  Host localhost:8080
6  Origin http://localhost:8081
7  …
8  …
```

这个请求将向服务端询问是否具备该资源的 DELETE 权限，服务端会给浏览器一个响应，代码如下：

```
1  …
2  …
3  HTTP/1.1 200
```

```
4   Access-Control-Allow-Origin: http://localhost:8081
5   Access-Control-Allow-Methods: DELETE
6   Access-Control-Max-Age: 1800
7   Allow: GET, HEAD, POST, PUT, DELETE, OPTIONS, PATCH
8   Content-Length: 0
9   Date: Thu, 12 Jul 2018 13:20:26 GMT
10  …
11  …
```

服务端给浏览器的响应，Allow 头信息表示服务端支持的请求方法，这个请求相当于一个探测请求，当浏览器分析了请求头字段之后，知道服务端支持本次请求，则进入第二步。

第二步：发送 DELETE 请求。接下来浏览器就会发送一个跨域的 DELETE 请求，代码如下：

```
1   …
2   …
3   Host: localhost:8080
4   Origin: http://localhost:8081
5   Connection: keep-alive
6   …
7   …
```

服务端给一个响应：

```
1   …
2   …
3   HTTP/1.1 200
4   Access-Control-Allow-Origin: http://localhost:8081
5   Content-Type: text/plain;charset=UTF-8
6   Date: Thu, 12 Jul 2018 13:20:26 GMT
7   …
8   …
```

至此，一个跨域的 DELETE 请求完成。

无论是简单请求还是需要先进行探测的请求，前端的写法都是不变的，额外的处理都是在服务端来完成的。在传统的 Java EE 开发中，可以通过过滤器统一配置，而 Spring Boot 中对此则提供了更加简洁的解决方案。在 Spring Boot 中配置 CORS 的步骤如下：

1. 创建 Spring Boot 工程

首先创建一个 Spring Boot 工程，添加 Web 依赖，代码如下：

```
1   <dependency>
2       <groupId>org.springframework.boot</groupId>
3       <artifactId>spring-boot-starter-web</artifactId>
4   </dependency>
```

2. 创建控制器

在新创建的 Spring Boot 工程中，添加一个 BookController 控制器，代码如下：

```
1   @RestController
2   @RequestMapping("/book")
```

```
3   public class BookController {
4       @PostMapping("/")
5       public String addBook(String name) {
6           return "receive:" + name;
7       }
8       @DeleteMapping("/{id}")
9       public String deleteBookById(@PathVariable Long id) {
10          return String.valueOf(id);
11      }
12  }
```

BookController 中提供了两个接口：一个是添加接口，另一个是删除接口。

3. 配置跨域

跨域有两个地方可以配置。一个是直接在相应的请求方法上加注解：

```
1   @RestController
2   @RequestMapping("/book")
3   public class BookController {
4       @PostMapping("/")
5       @CrossOrigin(value = "http://localhost:8081"
6               ,maxAge = 1800,allowedHeaders = "*")
7       public String addBook(String name) {
8           return "receive:" + name;
9       }
10      @DeleteMapping("/{id}")
11      @CrossOrigin(value = "http://localhost:8081"
12              ,maxAge = 1800,allowedHeaders = "*")
13      public String deleteBookById(@PathVariable Long id) {
14          return String.valueOf(id);
15      }
16  }
```

代码解释：

- @CrossOrigin 中的 value 表示支持的域，这里表示来自 http://localhost:8081 域的请求是支持跨域的。
- maxAge 表示探测请求的有效期，在前面的讲解中，读者已经了解到对于 DELETE、PUT 请求或者有自定义头信息的请求，在执行过程中会先发送探测请求，探测请求不用每次都发送，可以配置一个有效期，有效期过了之后才会发送探测请求。这个属性默认是 1800 秒，即 30 分钟。
- allowedHeaders 表示允许的请求头，*表示所有的请求头都被允许。

这种配置方式是一种细粒度的配置，可以控制到每一个方法上。当然，也可以不在每个方法上添加@CrossOrigin 注解，而是采用一种全局配置，代码如下：

```
1   @Configuration
2   public class MyWebMvcConfig implements WebMvcConfigurer {
3       @Override
```

```
4       public void addCorsMappings(CorsRegistry registry) {
5           registry.addMapping("/book/**")
6                   .allowedHeaders("*")
7                   .allowedMethods("*")
8                   .maxAge(1800)
9                   .allowedOrigins("http://localhost:8081");
10      }
11  }
```

代码解释：

- 全局配置需要自定义类实现 WebMvcConfigurer 接口，然后实现接口中的 addCorsMappings 方法。
- 在 addCorsMappings 方法中，addMapping 表示对哪种格式的请求路径进行跨域处理；allowedHeaders 表示允许的请求头，默认允许所有的请求头信息；allowedMethods 表示允许的请求方法，默认是 GET、POST 和 HEAD；*表示支持所有的请求方法；maxAge 表示探测请求的有效期；allowedOrigins 表示支持的域。

在上面的两种配置方式（@CrossOrigin 注解配置和全局配置）中，选择其中一种即可，然后启动项目。

4．测试

新建一个 Spring Boot 项目，添加 Web 依赖，然后在 resources/static 目录下加入 jquery.js，再在 resources/static 目录下创建一个 index.html 文件，内容如下：

```
1   <!DOCTYPE html>
2   <html lang="en">
3   <head>
4   <meta charset="UTF-8">
5   <title>Title</title>
6   <script src="jquery3.3.1.js"></script>
7   </head>
8   <body>
9   <div id="contentDiv"></div>
10  <div id="deleteResult"></div>
11  <input type="button" value="提交数据" onclick="getData()"><br>
12  <input type="button" value="删除数据" onclick="deleteData()"><br>
13  <script>
14      function deleteData() {
15          $.ajax({
16              url:'http://localhost:8080/book/99',
17              type:'delete',
18              success:function (msg) {
19                  $("#deleteResult").html(msg);
20              }
21          })
22      }
23      function getData() {
24          $.ajax({
```

```
25              url:'http://localhost:8080/book/',
26              type:'post',
27              data:{name:'三国演义'},
28              success:function (msg) {
29                  $("#contentDiv").html(msg);
30              }
31          })
32      }
33  </script>
34  </body>
35  </html>
```

两个普通的 Ajax 都发送了一个跨域请求。

然后将项目的端口修改为 8081，代码如下：

```
1  server.port=8081
```

启动项目，在浏览器中输入"http://localhost:8081/index.html"，查看页面，然后分别单击两个按钮，查看请求结果，如图 4-24 所示。

图 4-24

4.7 配置类与 XML 配置

Spring Boot 推荐使用 Java 来完成相关的配置工作。在项目中，不建议将所有的配置放在一个配置类中，可以根据不同的需求提供不同的配置类，例如专门处理 Spring Security 的配置类、提供 Bean 的配置类、Spring MVC 相关的配置类。这些配置类上都需要添加@Configuration 注解，@ComponentScan 注解会扫描所有的 Spring 组件，也包括@Configuration。@ComponentScan 注解在项目入口类的@Spring BootApplication 注解中已经提供，因此在实际项目中只需要按需提供相关配置类即可。

Spring Boot 中并不推荐使用 XML 配置，建议尽量用 Java 配置代替 XML 配置，本书中的案例都是以 Java 配置为主。如果开发者需要使用 XML 配置，只需在 resources 目录下提供配置文件，然后通过@ImportResource 加载配置文件即可。例如，有一个 Hello 类如下：

```
1  public class Hello {
2      public String sayHello(String name) {
```

```
3        return "hello " + name;
4    }
5 }
```

在 resources 目录下新建 beans.xml 文件配置该类：

```
1 <beans xmlns="http://www.springframework.org/schema/beans"
2        xmlns:xsi="http://www.w3.org/2001/XMLSchema-instance"
3        xsi:schemaLocation="http://www.springframework.org/schema/beans
4 http://www.springframework.org/schema/beans/spring-beans.xsd">
5 <bean class="org.sang.Hello" id="hello"/>
6 </beans>
```

然后创建 Beans 配置类，导入 XML 配置：

```
1 @Configuration
2 @ImportResource("classpath:beans.xml")
3 public class Beans {
4 }
```

最后在 Controller 中就可以直接导入 Hello 类使用了：

```
1 @RestController
2 public class HelloController {
3     @Autowired
4     Hello hello;
5     @GetMapping("/hello")
6     public String hello() {
7         return hello.sayHello("江南一点雨");
8     }
9 }
```

4.8 注册拦截器

Spring MVC 中提供了 AOP 风格的拦截器，拥有更加精细的拦截处理能力。Spring Boot 中拦截器的注册更加方便，步骤如下：

步骤01 创建一个 Spring Boot 项目，添加 spring-boot-starter-web 依赖。

步骤02 创建拦截器实现 HandlerInterceptor 接口，代码如下：

```
1 public class MyInterceptor1 implements HandlerInterceptor {
2     @Override
3     public boolean preHandle(HttpServletRequest request,
4                              HttpServletResponse response,
5                              Object handler){
6         System.out.println("MyInterceptor1>>>preHandle");
7         return true;
8     }
9     @Override
```

```
10      public void postHandle(HttpServletRequest request,
11                             HttpServletResponse response,
12                             Object handler,
13                             ModelAndView modelAndView){
14          System.out.println("MyInterceptor1>>>postHandle");
15      }
16      @Override
17      public void afterCompletion(HttpServletRequest request,
18                                  HttpServletResponse response,
19                                  Object handler,
20                                  Exception ex){
21          System.out.println("MyInterceptor1>>>afterCompletion");
22      }
23  }
```

拦截器中的方法将按 preHandle→Controller→postHandle→afterCompletion 的顺序执行。注意，只有 preHandle 方法返回 true 时后面的方法才会执行。当拦截器链内存在多个拦截器时，postHandler 在拦截器链内的所有拦截器返回成功时才会调用，而 afterCompletion 只有 preHandle 返回 true 才调用，但若拦截器链内的第一个拦截器的 preHandle 方法返回 false，则后面的方法都不会执行。

步骤03 配置拦截器。定义配置类进行拦截器的配置，代码如下：

```
1  @Configuration
2  public class WebMvcConfig implements WebMvcConfigurer {
3      @Override
4      public void addInterceptors(InterceptorRegistry registry) {
5          registry.addInterceptor(new MyInterceptor1())
6                  .addPathPatterns("/**")
7                  .excludePathPatterns("/hello");
8      }
9  }
```

自定义类实现 WebMvcConfigurer 接口，实现接口中的 addInterceptors 方法。其中，addPathPatterns 表示拦截路径，excludePathPatterns 表示排除的路径。

步骤04 测试。在浏览器中提供/hello2 和/hello 接口分别进行访问，当访问/hello2 接口时，打印日志，如图 4-25 所示。

```
2018-07-14 20:41:02.915  INFO 2964
2018-07-14 20:41:02.934  INFO 2964
MyInterceptor1>>>preHandle
MyInterceptor1>>>postHandle
MyInterceptor1>>>afterCompletion
```

图 4-25

4.9 启动系统任务

有一些特殊的任务需要在系统启动时执行，例如配置文件加载、数据库初始化等操作。如果没有使用 Spring Boot，这些问题可以在 Listener 中解决。Spring Boot 对此提供了两种解决方案：CommandLineRunner 和 ApplicationRunner。CommandLineRunner 和 ApplicationRunner 基本一致，差别主要体现在参数上。

4.9.1 CommandLineRunner

Spring Boot 项目在启动时会遍历所有 CommandLineRunner 的实现类并调用其中的 run 方法，如果整个系统中有多个 CommandLineRunner 的实现类，那么可以使用@Order 注解对这些实现类的调用顺序进行排序。

在一个 Spring Boot Web 项目中添加两个 CommandLineRunner，分别如下：

```
@Component
@Order(1)
public class MyCommandLineRunner1 implements CommandLineRunner {
    @Override
    public void run(String... args) throws Exception {
        System.out.println("Runner1>>>"+Arrays.toString(args));
    }
}
@Component
@Order(2)
public class MyCommandLineRunner2 implements CommandLineRunner {
    @Override
    public void run(String... args) throws Exception {
        System.out.println("Runner2>>>"+Arrays.toString(args));
    }
}
```

代码解释：

- @Order(1)注解用来描述 CommandLineRunner 的执行顺序，数字越小越先执行。
- run 方法中是调用的核心逻辑，参数是系统启动时传入的参数，即入口类中 main 方法的参数（在调用 SpringApplication.run 方法时被传入 Spring Boot 项目中）。

在系统启动时，配置传入的参数。以 IntelliJ IDEA 为例，配置方式如下：

步骤01 单击右上角的编辑启动配置，如图 4-26 所示。

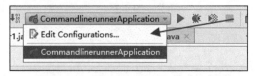

图 4-26

步骤02 在打开的新页面中编辑 Program arguments，如果有多个参数，参数之间用空格隔开，如图 4-27 所示。

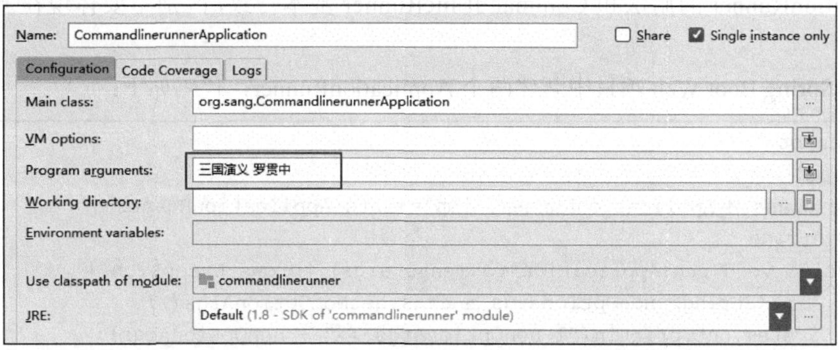

图 4-27

步骤03 启动项目，启动日志如图 4-28 所示。

```
2018-07-14 21:59:30.961  INFO 6764 --- [
Runner1>>>[三国演义，罗贯中]
Runner2>>>[三国演义，罗贯中]
```

图 4-28

在 Eclipse 中配置启动参数时，先选中启动类并右击，选择 Run As，再选择 Run Configurations，在新打开的页面中选择 Arguments 选项卡，填入相关参数（多个参数之间用空格隔开），如图 4-29 所示。

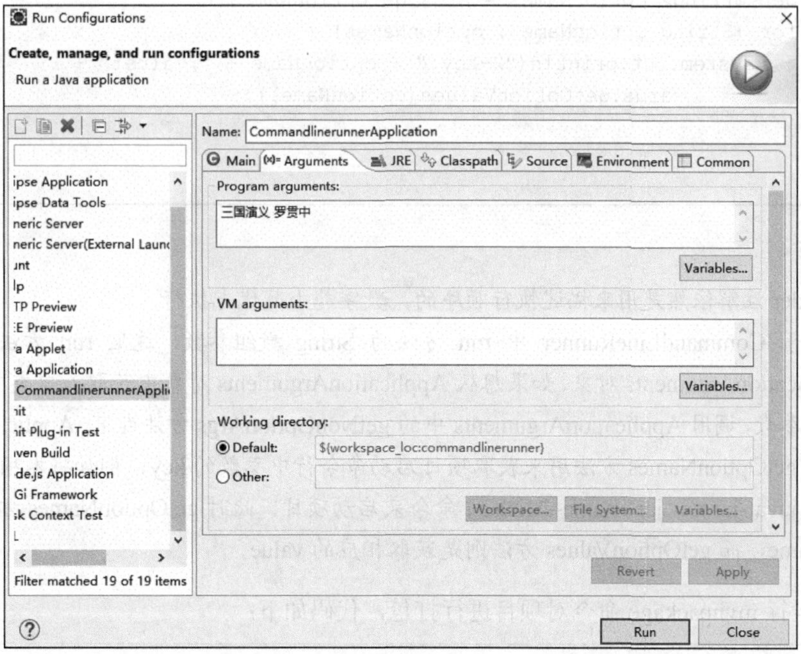

图 4-29

4.9.2 ApplicationRunner

ApplicationRunner 的用法和 CommandLineRunner 基本一致，区别主要体现在 run 方法的参数上。

在一个 Spring Boot Web 项目中新建两个 ApplicationRunner，代码如下：

```java
@Component
@Order(2)
public class MyApplicationRunner1 implements ApplicationRunner {
    @Override
    public void run(ApplicationArguments args) throws Exception {
        List<String> nonOptionArgs = args.getNonOptionArgs();
        System.out.println("1-nonOptionArgs>>>" + nonOptionArgs);
        Set<String> optionNames = args.getOptionNames();
        for (String optionName : optionNames) {
            System.out.println("1-key:" + optionName + ";value:" +
                    args.getOptionValues(optionName));
        }
    }
}
@Component
@Order(1)
public class MyApplicationRunner2 implements ApplicationRunner {
    @Override
    public void run(ApplicationArguments args) throws Exception {
        List<String> nonOptionArgs = args.getNonOptionArgs();
        System.out.println("2-nonOptionArgs>>>" + nonOptionArgs);
        Set<String> optionNames = args.getOptionNames();
        for (String optionName : optionNames) {
            System.out.println("2-key:" + optionName + ";value:" +
                    args.getOptionValues(optionName));
        }
    }
}
```

代码解释：

- @Order 注解依然是用来描述执行顺序的，数字越小越优先执行。
- 不同于 CommandLineRunner 中 run 方法的 String 数组参数，这里 run 方法的参数是一个 ApplicationArguments 对象，如果想从 ApplicationArguments 对象中获取入口类中 main 方法接收的参数，调用 ApplicationArguments 中的 getNonOptionArgs 方法即可。ApplicationArguments 中的 getOptionNames 方法用来获取项目启动命令行中参数的 key，例如将本项目打成 jar 包，运行 java -jar xxx.jar --name=Michael 命令来启动项目，此时 getOptionNames 方法获取到的就是 name，而 getOptionValues 方法则是获取相应的 value。

接下来运行 mvnpackage 命令对项目进行打包，代码如下：

```
mvn package
```

进入打包目录中，执行如下命令启动项目：

```
1  Java -jar linerunner-0.0.1.jar --name=Michael --age=99 三国演义 罗贯中
```

命令解释：

- --name=Michael --age=99 都属于 getOptionNames/getOptionValues 范畴。
- 后面的"三国演义""罗贯中"可以通过 getNonOptionArgs 方法获取，获取到的是一个数组，相当于上文提到的运行时配置的 ProgramArguments。

项目启动结果如图 4-30 所示。

```
2-nonOptionArgs>>>[三国演义, 罗贯中]
2-key:name;value:[Michael]
2-key:age;value:[99]
1-nonOptionArgs>>>[三国演义, 罗贯中]
1-key:name;value:[Michael]
1-key:age;value:[99]
```

图 4-30

4.10 整合 Servlet、Filter 和 Listener

一般情况下，使用 Spring、Spring MVC 这些框架之后，基本上就告别 Servlet、Filter 以及 Listener 了，但是有时在整合一些第三方框架时，可能还是不得不使用 Servlet，比如在整合某报表插件时就需要使用 Servlet。Spring Boot 中对于整合这些基本的 Web 组件也提供了很好的支持。

在一个 Spring Boot Web 项目中添加如下三个组件：

```
1   @WebServlet("/my")
2   public class MyServlet extends HttpServlet {
3       @Override
4       protected void doGet(HttpServletRequest req, HttpServletResponse resp){
5           doPost(req,resp);
6       }
7       @Override
8       protected void doPost(HttpServletRequest req, HttpServletResponse resp){
9           System.out.println("name>>>"+req.getParameter("name"));
10      }
11  }
12  @WebFilter("/*")
13  public class MyFilter implements Filter {
14      @Override
15      public void init(FilterConfig filterConfig){
16          System.out.println("MyFilter>>>init");
17      }
18      @Override
19      public void doFilter(ServletRequest req, ServletResponse resp, FilterChain
```

```
20  chain){
21      System.out.println("MyFilter>>>doFilter");
22      chain.doFilter(req,resp);
23  }
24  @Override
25  public void destroy() {
26      System.out.println("MyFilter>>>destroy");
27  }
28  }
29  @WebListener
30  public class MyListener implements ServletRequestListener {
31      @Override
32      public void requestDestroyed(ServletRequestEvent sre) {
33          System.out.println("MyListener>>>requestDestroyed");
34      }
35      @Override
36      public void requestInitialized(ServletRequestEvent sre) {
37          System.out.println("MyListener>>>requestInitialized");
38      }
    }
```

代码解释：

- 这里定义了三个基本的组件，分别使用@WebServlet、@WebFilter 和@WebListener 三个注解进行标记。
- 这里以 ServletRequestListener 为例，但是对于其他的 Listener，例如 HttpSessionListener、ServletContextListener 等也是支持的。

在项目入口类上添加@ServletComponentScan 注解，实现对 Servlet、Filter 以及 Listener 的扫描，代码如下：

```
1  @Spring BootApplication
2  @ServletComponentScan
3  public class ServletApplication {
4      public static void main(String[] args) {
5          SpringApplication.run(ServletApplication.class, args);
6      }
7  }
```

最后，启动项目，在浏览器中输入"http://localhost:8080/my?name=Michael"，可以看到相关日志，如图 4-31 所示。

```
2018-07-14 23:39:06.631   INFO 9204 —
2018-07-14 23:39:06.631   INFO 9204 —
MyFilter>>>init
2018-07-14 23:39:06.792   INFO 9204 —
2018-07-14 23:39:07.062   INFO 9204 —
2018-07-14 23:39:07.121   INFO 9204 —
2018-07-14 23:39:07.137   INFO 9204 —
2018-07-14 23:39:07.153   INFO 9204 —
2018-07-14 23:39:07.153   INFO 9204 —
2018-07-14 23:39:07.405   INFO 9204 —
2018-07-14 23:39:07.479   INFO 9204 —
2018-07-14 23:39:07.484   INFO 9204 —
MyListener>>>requestInitialized
MyFilter>>>doFilter
name>>>Michael
MyListener>>>requestDestroyed
```

图 4-31

4.11　路径映射

一般情况下，使用了页面模板后，用户需要通过控制器才能访问页面。有一些页面需要在控制器中加载数据，然后渲染，才能显示出来；还有一些页面在控制器中不需要加载数据，只是完成简单的跳转，对于这种页面，可以直接配置路径映射，提高访问速度。例如，有两个 Thymeleaf 做模板的页面 login.html 和 index.html，直接在 MVC 配置中重写 addViewControllers 方法配置映射关系即可：

```
1  @Configuration
2  public class WebMvcConfig implements WebMvcConfigurer {
3      @Override
4      public void addViewControllers(ViewControllerRegistry registry) {
5          registry.addViewController("/login").setViewName("login");
6          registry.addViewController("/index").setViewName("index");
7      }
8  }
```

配置完成后，就可以直接访问 http://localhost:8080/login 等地址了。

4.12　配置 AOP

4.12.1　AOP 简介

要介绍面向切面编程（Aspect-Oriented Programming，AOP），需要读者首先考虑这样一个场景：公司有一个人力资源管理系统目前已经上线，但是系统运行不稳定，有时运行得很慢，为了检测出到底是哪个环节出问题了，开发人员想要监控每一个方法的执行时间，再根据这些执行时间判断出问题所在。当问题解决后，再把这些监控移除掉。系统目前已经运行，如果手动修改系统中成千上万个方法，那么工作量未免太大，而且这些监控方法以后还要移除掉；如果能够在系统运行过程中动态添加代码，就能很好地解决这个需求。这种在系统运行时动态添加代码的方式称为面向切面编程（AOP）。Spring 框架对 AOP 提供了很好的支持。在 AOP 中，有一些常见的概念需要读者了解。

- Joinpoint（连接点）：类里面可以被增强的方法即为连接点。例如，想修改哪个方法的功能，那么该方法就是一个连接点。
- Pointcut（切入点）：对 Joinpoint 进行拦截的定义即为切入点。例如，拦截所有以 insert 开始的方法，这个定义即为切入点。
- Advice（通知）：拦截到 Joinpoint 之后所要做的事情就是通知。例如，上文说到的打印日志监控。通知分为前置通知、后置通知、异常通知、最终通知和环绕通知。
- Aspect（切面）：Pointcut 和 Advice 的结合。
- Target（目标对象）：要增强的类称为 Target。

这些是对 AOP 的简单介绍，接下来看看如何在 Spring Boot 中实现 AOP。

4.12.2 Spring Boot 支持

Spring Boot 在 Spring 的基础上对 AOP 的配置提供了自动化配置解决方案 spring-boot-starter-aop，使开发者能够更加便捷地在 Spring Boot 项目中使用 AOP。配置步骤如下。

首先在 Spring Boot Web 项目中引入 spring-boot-starter-aop 依赖，代码如下：

```
1  <dependency>
2      <groupId>org.springframework.boot</groupId>
3      <artifactId>spring-boot-starter-aop</artifactId>
4  </dependency>
```

然后在 org.sang.aop.service 包下创建 UserService 类，代码如下：

```
1   @Service
2   public class UserService {
3       public String getUserById(Integer id) {
4           System.out.println("get...");
5           return "user";
6       }
7       public void deleteUserById(Integer id) {
8           System.out.println("delete...");
9       }
10  }
```

接下来创建切面，代码如下：

```
1   @Component
```

```
 2    @Aspect
 3    public class LogAspect {
 4        @Pointcut("execution(* org.sang.aop.service.*.*(..))")
 5        public void pc1() {
 6        }
 7        @Before(value = "pc1()")
 8        public void before(JoinPoint jp) {
 9            String name = jp.getSignature().getName();
10            System.out.println(name + "方法开始执行...");
11        }
12        @After(value = "pc1()")
13        public void after(JoinPoint jp) {
14            String name = jp.getSignature().getName();
15            System.out.println(name + "方法执行结束...");
16        }
17        @AfterReturning(value = "pc1()", returning = "result")
18        public void afterReturning(JoinPoint jp, Object result) {
19            String name = jp.getSignature().getName();
20            System.out.println(name + "方法返回值为：" + result);
21        }
22        @AfterThrowing(value = "pc1()",throwing = "e")
23        public void afterThrowing(JoinPoint jp,Exception e) {
24            String name = jp.getSignature().getName();
25            System.out.println(name+"方法抛异常了,异常是："+e.getMessage());
26        }
27        @Around("pc1()")
28        public Object around(ProceedingJoinPoint pjp) throws Throwable {
29            return pjp.proceed();
30        }
31    }
```

代码解释：

- @Aspect 注解表明这是一个切面类。
- 第 4~6 行定义的 pc1 方法使用了 @Pointcut 注解，这是一个切入点定义。execution 中的第一个*表示方法返回任意值，第二个*表示 service 包下的任意类，第三个*表示类中的任意方法，括号中的两个点表示方法参数任意，即这里描述的切入点为 service 包下所有类中的所有方法。
- 第 7~11 行定义的方法使用了 @Before 注解，表示这是一个前置通知，该方法在目标方法执行之前执行。通过 JoinPoint 参数可以获取目标方法的方法名、修饰符等信息。
- 第 12~16 行定义的方法使用了 @After 注解，表示这是一个后置通知，该方法在目标方法执行之后执行。
- 第 17~21 行定义的方法使用了 @AfterReturning 注解，表示这是一个返回通知，在该方法中可以获取目标方法的返回值。@AfterReturning 注解的 returning 参数是指返回值的变量名，对应方法的参数。注意，在方法参数中定义了 result 的类型为 Object，表示目标方法的返回值可以是任意类型，若 result 参数的类型为 Long，则该方法只能处理目标方法返回值为 Long 的情况。

- 第22~26 行定义的方法使用了@AfterThrowing 注解,表示这是一个异常通知,即当目标方法发生异常时,该方法会被调用,异常类型为 Exception 表示所有的异常都会进入该方法中执行,若异常类型为 ArithmeticException,则表示只有目标方法抛出的 ArithmeticException 异常才会进入该方法中处理。
- 第27~30 行定义的方法使用了@Around 注解,表示这是一个环绕通知。环绕通知是所有通知里功能最为强大的通知,可以实现前置通知、后置通知、异常通知以及返回通知的功能。目标方法进入环绕通知后,通过调用 ProceedingJoinPoint 对象的 proceed 方法使目标方法继续执行,开发者可以在此修改目标方法的执行参数、返回值等,并且可以在此处理目标方法的异常。

配置完成后,接下来在 Controller 中创建接口,分别调用 UserService 中的两个方法,即可看到 LogAspect 中的代码动态地嵌入目标方法中执行了。UserController 类的定义如下:

```
1  @RestController
2  public class UserController {
3      @Autowired
4      UserService userService;
5      @GetMapping("/getUserById")
6      public String getUserById(Integer id) {
7          return userService.getUserById(id);
8      }
9      @GetMapping("/deleteUserById")
10     public void deleteUserById(Integer id) {
11         userService.deleteUserById(id);
12     }
13 }
```

4.13 其 他

4.13.1 自定义欢迎页

Spring Boot 项目在启动后,首先会去静态资源路径下查找 index.html 作为首页文件,若查找不到,则会去查找动态的 index 文件作为首页文件。

例如,如果想使用静态的 index.html 页面作为项目首页,那么只需在 resources/static 目录下创建 index.html 文件即可。若想使用动态页面作为项目首页,则需在 resources/templates 目录下创建 index.html(使用 Thymeleaf 模板)或者 index.ftl(使用 FreeMarker 模板),然后在 Controller 中返回逻辑视图名,代码如下:

```
1  @RequestMapping("/index")
2  public String hello() {
3      return "index";
4  }
```

最后启动项目，输入"http://localhost:8080/"就可以看到项目首页的内容了。

4.13.2 自定义 favicon

favicon.ico 是浏览器选项卡左上角的图标，可以放在静态资源路径下或者类路径下，静态资源路径下的 favicon.ico 优先级高于类路径下的 favicon.ico。

可以使用在线转换网站 https://jinaconvert.com/cn/convert-to-ico.php 将一张普通图片转为 .ico 图片，转换成功后，将文件重命名为 favicon.ico，然后复制到 resources/static 目录下，如图 4-32 所示。

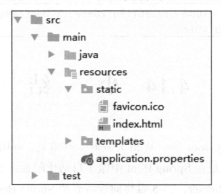

图 4-32

最后启动项目，就可以在浏览器选项卡中看到效果了，如图 4-33 所示。

图 4-33

4.13.3 除去某个自动配置

Spring Boot 中提供了大量的自动化配置类，例如上文提到过的 ErrorMvcAutoConfiguration、ThymeleafAutoConfiguration、FreeMarkerAutoConfiguration、MultipartAutoConfiguration 等，这些自动化配置可以减少相应操作的配置，达到开箱即用的效果。在 Spring Boot 的入口类上有一个 @Spring BootApplication 注解。该注解是一个组合注解，由 @Spring BootConfiguration、@EnableAutoConfiguration 以及 @ComponentScan 组成，其中 @EnableAutoConfiguration 注解开启自动化配置，相关的自动化配置类就会被使用。如果开发者不想使用某个自动化配置，按如下方式除去相关配置即可：

```
1  @Spring BootApplication
2  @EnableAutoConfiguration(exclude = {ErrorMvcAutoConfiguration.class})
3  public class OtherApplication {
4    public static void main(String[] args) {
```

```
5            SpringApplication.run(OtherApplication.class, args);
6        }
7    }
```

在@EnableAutoConfiguration 注解中使用 exclude 属性除去 Error 的自动化配置类，这时如果在 resources/static/error 目录下创建 4xx.html、5xx.html（具体参见 4.5 节），访问出错时就不会自动跳转了。由于@EnableAutoConfiguration 注解的 exclude 属性值是一个数组，因此有多个要排除的自动化配置类时只需继续添加即可。除了这种配置方式外，开发者也可以在 application.properties 配置文件中进行配置，代码如下：

```
1    spring.autoconfigure.exclude=org.springframework.boot.autoconfigure.web.servlet.error.ErrorMvcAutoConfiguration
```

4.14 小　　结

本章向读者介绍了 Spring Boot 整合 Web 开发时一些常见、有用的配置。在这些配置中，大部分是 Spring MVC 的功能，只是在 Spring Boot 中做了自动化配置，少部分是 Spring Boot 自身提供的功能，例如 CommandLineRunner。第 5 章将向读者介绍 Spring Boot 整合持久层技术。

第 5 章

Spring Boot 整合持久层技术

本章概要

- 整合 JdbcTemplate
- 整合 MyBatis
- 整合 Spring Data JPA
- 多数据源

持久层是 Java EE 中访问数据库的核心操作，Spring Boot 中对常见的持久层框架都提供了自动化配置，例如 JdbcTemplate、JPA 等，MyBatis 的自动化配置则是 MyBatis 官方提供的。接下来分别向读者介绍 Spring Boot 整合这几种持久层技术。

5.1 整合 JdbcTemplate

JdbcTemplate 是 Spring 提供的一套 JDBC 模板框架，利用 AOP 技术来解决直接使用 JDBC 时大量重复代码的问题。JdbcTemplate 虽然没有 MyBatis 那么灵活，但是比直接使用 JDBC 要方便很多。Spring Boot 中对 JdbcTemplate 的使用提供了自动化配置类 JdbcTemplateAutoConfiguration，部分源码如下：

```
1  @Configuration
2  @ConditionalOnClass({ DataSource.class, JdbcTemplate.class })
3  @ConditionalOnSingleCandidate(DataSource.class)
4  @AutoConfigureAfter(DataSourceAutoConfiguration.class)
```

```
5    @EnableConfigurationProperties(JdbcProperties.class)
6    public class JdbcTemplateAutoConfiguration {
7      @Configuration
8      static class JdbcTemplateConfiguration {
9        private final DataSource dataSource;
10       private final JdbcProperties properties;
11       JdbcTemplateConfiguration(DataSource dataSource, JdbcProperties properties) {
12         this.dataSource = dataSource;
13         this.properties = properties;
14       }
15       @Bean
16       @Primary
17       @ConditionalOnMissingBean(JdbcOperations.class)
18       public JdbcTemplate jdbcTemplate() {
19         JdbcTemplate jdbcTemplate = new JdbcTemplate(this.dataSource);
20         JdbcProperties.Template template = this.properties.getTemplate();
21         jdbcTemplate.setFetchSize(template.getFetchSize());
22         jdbcTemplate.setMaxRows(template.getMaxRows());
23         if (template.getQueryTimeout() != null) {
24           jdbcTemplate
25               .setQueryTimeout((int) template
26   .getQueryTimeout()
27   .getSeconds());
28         }
29         return jdbcTemplate;
30       }
31     }
32   ...
33   ...
34   }
```

从上面这段源码中可以看出,当 classpath 下存在 DataSource 和 JdbcTemplate 并且 DataSource 只有一个实例时,自动配置才会生效,若开发者没有提供 JdbcOperations,则 Spring Boot 会自动向容器中注入一个 JdbcTemplate(JdbcTemplate 是 JdbcOperations 的子类)。由此可以看到,开发者想要使用 JdbcTemplate,只需要提供 JdbcTemplate 的依赖和 DataSource 依赖即可。具体操作步骤如下。

1. 创建数据库和表

在数据库中创建表,代码如下:

```
1  CREATE DATABASE `chapter05` DEFAULT CHARACTER SET utf8;
2  USE `chapter05`;
3  CREATE TABLE `book` (
4    `id` int(11) NOT NULL AUTO_INCREMENT,
5    `name` varchar(128) DEFAULT NULL,
6    `author` varchar(64) DEFAULT NULL,
7    PRIMARY KEY (`id`)
8  ) ENGINE=InnoDB DEFAULT CHARSET=utf8;
9  insert into `book`(`id`,`name`,`author`) values (1,'三国演义','罗贯中'),(2,'水浒传','
```

施耐庵');

创建 chapter05 数据库,在库中创建 book 表,同时添加两条测试语句。

2. 创建项目

创建 Spring Boot 项目,添加如下依赖:

```
1  <dependency>
2      <groupId>org.springframework.boot</groupId>
3      <artifactId>spring-boot-starter-jdbc</artifactId>
4  </dependency>
5  <dependency>
6      <groupId>org.springframework.boot</groupId>
7      <artifactId>spring-boot-starter-web</artifactId>
8  </dependency>
9  <dependency>
10     <groupId>mysql</groupId>
11     <artifactId>mysql-connector-java</artifactId>
12     <scope>runtime</scope>
13 </dependency>
14 <dependency>
15     <groupId>com.alibaba</groupId>
16     <artifactId>druid</artifactId>
17     <version>1.1.9</version>
18 </dependency>
```

spring-boot-starter-jdbc 中提供了 spring-jdbc,另外还加入了数据库驱动依赖和数据库连接池依赖。

3. 数据库配置

在 application.properties 中配置数据库基本连接信息:

```
1  spring.datasource.type=com.alibaba.druid.pool.DruidDataSource
2  spring.datasource.url=jdbc:mysql:///chapter05
3  spring.datasource.username=root
4  spring.datasource.password=123
```

4. 创建实体类

创建 Book 实体类,代码如下:

```
1  public class Book {
2      private Integer id;
3      private String name;
4      private String author;
5      //省略 getter/setter
6  }
```

5. 创建数据库访问层

创建 BookDao,代码如下:

```
1  @Repository
```

```java
public class BookDao {
    @Autowired
    JdbcTemplate jdbcTemplate;
    public int addBook(Book book) {
        return jdbcTemplate.update("INSERT INTO book(name,author) VALUES (?,?)",
                book.getName(), book.getAuthor());
    }
    public int updateBook(Book book) {
        return jdbcTemplate.update("UPDATE book SET name=?,author=? WHERE id=?",
                book.getName(), book.getAuthor(), book.getId());
    }
    public int deleteBookById(Integer id) {
        return jdbcTemplate.update("DELETE FROM book WHERE id=?", id);
    }
    public Book getBookById(Integer id) {
        return jdbcTemplate.queryForObject("select * from book where id=?",
                new BeanPropertyRowMapper<>(Book.class), id);
    }
    public List<Book> getAllBooks() {
        return jdbcTemplate.query("select * from book",
                new BeanPropertyRowMapper<>(Book.class));
    }
}
```

代码解释：

- 创建 BookDao，注入 JdbcTemplate。由于已经添加了 spring-jdbc 相关的依赖，JdbcTemplate 会被自动注册到 Spring 容器中，因此这里可以直接注入 JdbcTemplate 使用。
- 在 JdbcTemplate 中，增删改三种类型的操作主要使用 update 和 batchUpdate 方法来完成。query 和 queryForObject 方法主要用来完成查询功能。另外，还有 execute 方法可以用来执行任意的 SQL、call 方法用来调用存储过程等。
- 在执行查询操作时，需要有一个 RowMapper 将查询出来的列和实体类中的属性一一对应起来。如果列名和属性名都是相同的，那么可以直接使用 BeanPropertyRowMapper；如果列名和属性名不同，就需要开发者自己实现 RowMapper 接口，将列和实体类属性一一对应起来。

6. 创建 Service 和 Controller

创建 BookService 和 BookController，代码如下：

```java
@Service
public class BookService {
    @Autowired
    BookDao bookDao;
    public int addBook(Book book) {
        return bookDao.addBook(book);
    }
    public int updateBook(Book book) {
        return bookDao.updateBook(book);
    }
    public int deleteBookById(Integer id) {
```

```
12          return bookDao.deleteBookById(id);
13      }
14      public Book getBookById(Integer id) {
15          return bookDao.getBookById(id);
16      }
17      public List<Book> getAllBooks() {
18          return bookDao.getAllBooks();
19      }
20  }
21  @RestController
22  public class BookController {
23      @Autowired
24      BookService bookService;
25      @GetMapping("/bookOps")
26      public void bookOps() {
27          Book b1 = new Book();
28          b1.setName("西厢记");
29          b1.setAuthor("王实甫");
30          int i = bookService.addBook(b1);
31          System.out.println("addBook>>>" + i);
32          Book b2 = new Book();
33          b2.setId(1);
34          b2.setName("朝花夕拾");
35          b2.setAuthor("鲁迅");
36          int updateBook = bookService.updateBook(b2);
37          System.out.println("updateBook>>>"+updateBook);
38          Book b3 = bookService.getBookById(1);
39          System.out.println("getBookById>>>"+b3);
40          int delete = bookService.deleteBookById(2);
41          System.out.println("deleteBookById>>>"+delete);
42          List<Book> allBooks = bookService.getAllBooks();
43          System.out.println("getAllBooks>>>"+allBooks);
44      }
45  }
```

最后,在浏览器中访问 http://localhost:8080/bookOps 地址,控制台打印日志如图 5-1 所示。

```
addBook>>>1
updateBook>>>1
getBookById>>>Book{id=1, name='朝花夕拾', author='鲁迅'}
deleteBookById>>>1
getAllBooks>>>[Book{id=1, name='朝花夕拾', author='鲁迅'}, Book{id=4, name='西厢记', author='王实甫'}]
```

图 5-1

数据库中的数据如图 5-2 所示。

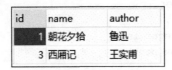

图 5-2

5.2 整合 MyBatis

MyBatis 是一款优秀的持久层框架，原名叫作 iBaits，2010 年由 ApacheSoftwareFoundation 迁移到 Google Code 并改名为 MyBatis，2013 年又迁移到 GitHub 上。MyBatis 支持定制化 SQL、存储过程以及高级映射。MyBatis 几乎避免了所有的 JDBC 代码手动设置参数以及获取结果集。在传统的 SSM 框架整合中，使用 MyBatis 需要大量的 XML 配置，而在 Spring Boot 中，MyBatis 官方提供了一套自动化配置方案，可以做到 MyBatis 开箱即用。具体使用步骤如下。

1. 创建项目

创建 Spring Boot 项目，添加 MyBatis 依赖、数据库驱动依赖以及数据库连接池依赖，代码如下：

```
1  <dependency>
2    <groupId>org.springframework.boot</groupId>
3    <artifactId>spring-boot-starter-web</artifactId>
4  </dependency>
5  <dependency>
6    <groupId>org.mybatis.spring.boot</groupId>
7    <artifactId>mybatis-spring-boot-starter</artifactId>
8    <version>1.3.2</version>
9  </dependency>
10 <dependency>
11   <groupId>com.alibaba</groupId>
12   <artifactId>druid</artifactId>
13   <version>1.1.9</version>
14 </dependency>
15 <dependency>
16   <groupId>mysql</groupId>
17   <artifactId>mysql-connector-java</artifactId>
18   <scope>runtime</scope>
19 </dependency>
```

2. 创建数据库、表、实体类等

数据库和表、实体类以及 application.properties 中配置的数据库连接信息都与上一节一致，这里不再赘述。

3. 创建数据库访问层

创建 BookMapper，代码如下：

```
1  @Mapper
2  public interface BookMapper {
3      int addBook(Book book);
4      int deleteBookById(Integer id);
5      int updateBookById(Book book);
6      Book getBookById();
7      List<Book> getAllBooks();
```

```
8  }
```

代码解释:

- 在项目的根包下面创建一个子包 Mapper, 在 Mapper 中创建 BookMapper。
- 有两种方式指明该类是一个 Mapper: 第一种如前面的代码所示, 在 BookMapper 上添加 @Mapper 注解, 表明该接口是一个 MyBatis 中的 Mapper, 这种方式需要在每一个 Mapper 上都添加注解; 还有一种简单的方式是在配置类上添加@MapperScan("org.sang.mapper")注解, 表示扫描 org.sang.mapper 包下的所有接口作为 Mapper, 这样就不需要在每个接口上配置 @Mapper 注解了。

4. 创建 BookMapper.xml

在与 BookMapper 相同的位置创建 BookMapper.xml 文件, 代码如下:

```
1   <?xml version="1.0" encoding="UTF-8" ?>
2   <!DOCTYPE mapper
3           PUBLIC "-//mybatis.org//DTD Mapper 3.0//EN"
4   "http://mybatis.org/dtd/mybatis-3-mapper.dtd">
5   <mapper namespace="org.sang.mapper.BookMapper">
6   <insert id="addBook" parameterType="org.sang.model.Book">
7       INSERT INTO book(name,author) VALUES (#{name},#{author})
8   </insert>
9   <delete id="deleteBookById" parameterType="int">
10          DELETE FROM book WHERE id=#{id}
11  </delete>
12  <update id="updateBookById" parameterType="org.sang.model.Book">
13          UPDATE book set name=#{name},author=#{author} WHERE id=#{id}
14  </update>
15  <select id="getBookById" parameterType="int" resultType="org.sang.model.Book">
16          SELECT * FROM book WHERE id=#{id}
17  </select>
18  <select id="getAllBooks" resultType="org.sang.model.Book">
19          SELECT * FROM book
20  </select>
21  </mapper>
```

代码解释:

- 针对 BookMapper 接口中的每一个方法都在 BookMapper.xml 中列出了实现。
- #{}用来代替接口中的参数, 实体类中的属性可以直接通过#{实体类属性名}获取。

5. 创建 Service 和 Controller

创建 BookService 与 BookController, 代码如下:

```
1   @Service
2   public class BookService {
3       @Autowired
4       BookMapper bookMapper;
5       public int addBook(Book book) {
```

```
6            return bookMapper.addBook(book);
7        }
8        public int updateBook(Book book) {
9            return bookMapper.updateBookById(book);
10       }
11       public int deleteBookById(Integer id) {
12           return bookMapper.deleteBookById(id);
13       }
14       public Book getBookById(Integer id) {
15           return bookMapper.getBookById(id);
16       }
17       public List<Book> getAllBooks() {
18           return bookMapper.getAllBooks();
19       }
20   }
21   @RestController
22   public class BookController {
23       @Autowired
24       BookService bookService;
25       @GetMapping("/bookOps")
26       public void bookOps() {
27           Book b1 = new Book();
28           b1.setName("西厢记");
29           b1.setAuthor("王实甫");
30           int i = bookService.addBook(b1);
31           System.out.println("addBook>>>" + i);
32           Book b2 = new Book();
33           b2.setId(1);
34           b2.setName("朝花夕拾");
35           b2.setAuthor("鲁迅");
36           int updateBook = bookService.updateBook(b2);
37           System.out.println("updateBook>>>"+updateBook);
38           Book b3 = bookService.getBookById(1);
39           System.out.println("getBookById>>>"+b3);
40           int delete = bookService.deleteBookById(2);
41           System.out.println("deleteBookById>>>"+delete);
42           List<Book> allBooks = bookService.getAllBooks();
43           System.out.println("getAllBooks>>>"+allBooks);
44       }
45   }
```

6. 配置 pom.xml 文件

在 Maven 工程中,XML 配置文件建议写在 resources 目录下,但是上文的 Mapper.xml 文件写在包下,Maven 在运行时会忽略包下的 XML 文件,因此需要在 pom.xml 文件中重新指明资源文件位置,配置如下:

```
1   <build>
2     <resources>
3       <resource>
4         <directory>src/main/java</directory>
```

```
5    <includes>
6      <include>**/*.xml</include>
7    </includes>
8   </resource>
9   <resource>
10    <directory>src/main/resources</directory>
11  </resource>
12 </resources>
13</build>
```

接下来在浏览器中输入"http://localhost:8080/bookOps",即可看到数据库中数据的变化,控制台也打印出相应的日志,如图 5-3 所示。

```
addBook>>>1
updateBook>>>1
getBookById>>>Book{id=1, name='朝花夕拾', author='鲁迅'}
deleteBookById>>>0
getAllBooks>>>[Book{id=1, name='朝花夕拾', author='鲁迅'}, Book{id=3, name='西厢记', author='王实甫'}, Book{id=5, name='西厢记', author='王实甫'}]
```

图 5-3

通过上面的例子可以看到,MyBatis 基本上实现了开箱即用的特性。自动化配置将开发者从繁杂的配置文件中解脱出来,以专注于业务逻辑的开发。

5.3 整合 Spring Data JPA

JPA(Java Persistence API)和 Spring Data 是两个范畴的概念。

作为一名 Java EE 工程师,基本都有听说过 Hibernate 框架。Hibernate 是一个 ORM 框架,而 JPA 则是一种 ORM 规范,JPA 和 Hibernate 的关系就像 JDBC 与 JDBC 驱动的关系,即 JPA 制定了 ORM 规范,而 Hibernate 是这些规范的实现(事实上,是先有 Hibernate 后有 JPA,JPA 规范的起草者也是 Hibernate 的作者),因此从功能上来说,JPA 相当于 Hibernate 的一个子集。

Spring Data 是 Spring 的一个子项目,致力于简化数据库访问,通过规范的方法名称来分析开发者的意图,进而减少数据库访问层的代码量。Spring Data 不仅支持关系型数据库,也支持非关系型数据库。Spring Data JPA 可以有效简化关系型数据库访问代码。

Spring Boot 整合 Spring Data JPA 的步骤如下。

1. 创建数据库

创建数据库 jpa,代码如下:

```
1  CREATE DATABASE `jpa` DEFAULT CHARACTER SET utf8;
```

> **注 意**
>
> 创建数据库即可,不用创建表。

2. 创建项目

创建 Spring Boot 项目，添加 MySQL 和 Spring Data JPA 的依赖，代码如下：

```xml
<dependency>
    <groupId>org.springframework.boot</groupId>
    <artifactId>spring-boot-starter-data-jpa</artifactId>
</dependency>
<dependency>
    <groupId>org.springframework.boot</groupId>
    <artifactId>spring-boot-starter-web</artifactId>
</dependency>
<dependency>
    <groupId>com.alibaba</groupId>
    <artifactId>druid</artifactId>
    <version>1.1.9</version>
</dependency>
<dependency>
    <groupId>mysql</groupId>
    <artifactId>mysql-connector-java</artifactId>
    <scope>runtime</scope>
</dependency>
```

3. 数据库配置

在 application.properties 中配置数据库基本信息以及 JPA 相关配置：

```properties
spring.datasource.type=com.alibaba.druid.pool.DruidDataSource
spring.datasource.url=jdbc:mysql:///jpa
spring.datasource.username=root
spring.datasource.password=123
spring.jpa.show-sql=true
spring.jpa.database=mysql
spring.jpa.hibernate.ddl-auto=update
spring.jpa.properties.hibernate.dialect=org.hibernate.dialect.MySQL57Dialect
```

这里的配置信息主要分为两大类，第 1~4 行是数据库基本信息配置，第 5~8 行是 JPA 相关配置。其中，第 5 行表示是否在控制台打印 JPA 执行过程生成的 SQL，第 6 行表示 JPA 对应的数据库是 MySQL，第 7 行表示在项目启动时根据实体类更新数据库中的表（其他可选值有 create、create-drop、validate、no），第 8 行则表示使用的数据库方言是 MySQL57Dialect。

4. 创建实体类

创建 Book 实体类，代码如下：

```java
@Entity(name = "t_book")
public class Book {
    @Id
    @GeneratedValue(strategy = GenerationType.IDENTITY)
    private Integer id;
    @Column(name = "book_name",nullable = false)
    private String name;
```

```
8        private String author;
9        private Float price;
10   @Transient
11   private String description;
12       //省略 getter/setter
13   }
```

代码解释：

- @Entity 注解表示该类是一个实体类，在项目启动时会根据该类自动生成一张表，表的名称即@Entity 注解中 name 的值，如果不配置 name，默认表名为类名。
- 所有的实体类都要有主键，@Id 注解表示该属性是一个主键，@GeneratedValue 注解表示主键自动生成，strategy 则表示主键的生成策略。
- 默认情况下，生成的表中字段的名称就是实体类中属性的名称，通过@Column 注解可以定制生成的字段的属性，name 表示该属性对应的数据表中字段的名称，nullable 表示该字段非空。
- @Transient 注解表示在生成数据库中的表时，该属性被忽略，即不生成对应的字段。

5. 创建 BookDao 接口

创建 BookDao 接口，继承 JpaRepository，代码如下：

```
1    public interface BookDao extends JpaRepository<Book,Integer>{
2        List<Book> getBooksByAuthorStartingWith(String author);
3        List<Book> getBooksByPriceGreaterThan(Float price);
4        @Query(value = "select * from t_book where id=(select max(id) from t_book)",
5    nativeQuery = true)
6    Book getMaxIdBook();
7    @Query("select b from t_book b where b.id>:id and b.author=:author")
8    List<Book> getBookByIdAndAuthor(@Param("author") String author,
9    @Param("id") Integer id);
10   @Query("select b from t_book b where b.id<?2 and b.name like %?1%")
11   List<Book> getBooksByIdAndName(String name, Integer id);
12   }
```

代码解释：

- 自定义 BookDao 继承自 JpaRepository。JpaRepository 中提供了一些基本的数据操作方法，有基本的增删改查、分页查询、排序查询等。
- 第 2 行定义的方法表示查询以某个字符开始的所有书。
- 第 3 行定义的方法表示查询单价大于某个值的所有书。
- 在 Spring Data JPA 中，只要方法的定义符合既定规范，Spring Data 就能分析出开发者的意图，从而避免开发者定义 SQL。所谓的既定规范，就是一定的方法命名规则。支持的命名规则如表 5-1 所示。

表 5-1 支持的命名规则

KeyWords	方法命名举例	对应的 SQL
And	findByNameAndAge	where name= ? and age =?
Or	findByNameOrAge	where name= ? or age=?

（续表）

KeyWords	方法命名举例	对应的 SQL
Is	findByAgeIs	where age= ?
Equals	findByIdEquals	where id= ?
Between	findByAgeBetween	where age between ? and ?
LessThan	findByAgeLessThan	where age < ?
LessThanEquals	findByAgeLessThanEquals	where age <= ?
GreaterThan	findByAgeGreaterThan	where age > ?
GreaterThanEquals	findByAgeGreaterThanEquals	where age >= ?
After	findByAgeAfter	where age > ?
Before	findByAgeBefore	where age < ?
IsNull	findByNameIsNull	where name is null
isNotNull,NotNull	findByNameNotNull	where name is not null
Not	findByGenderNot	where gender <> ?
In	findByAgeIn	where age in (?)
NotIn	findByAgeNotIn	where age not in (?)
NotLike	findByNameNotLike	where name not like ?
Like	findByNameLike	where name like ?
StartingWith	findByNameStartingWith	where name like '?%'
EndingWith	findByNameEndingWith	where name like '%?'
Containing,Contains	findByNameContaining	where name like '%?%'
OrderBy	findByAgeGreaterThanOrderByIdDesc	where age>? order by id desc
True	findByEnabledTue	where enabled= true
False	findByEnabledFalse	where enabled= false
IgnoreCase	findByNameIgnoreCase	where UPPER(name)=UPPER(?)

- 既定的方法命名规则不一定满足所有的开发需求，因此 Spring Data JPA 也支持自定义 JPQL（Java Persistence Query Language）或者原生 SQL。第 4~6 行表示查询 id 最大的书，nativeQuery = true 表示使用原生的 SQL 查询。
- 第 7~9 行表示根据 id 和 author 进行查询，这里使用默认的 JPQL 语句。JPQL 是一种面向对象表达式语言，可以将 SQL 语法和简单查询语义绑定在一起，使用这种语言编写的查询是可移植的，可以被编译成所有主流数据库服务器上的 SQL。JPQL 与原生 SQL 语句类似，并且完全面向对象，通过类名和属性访问，而不是表名和表的属性（用过 Hibernate 的读者会觉得这类似于 HQL）。第 7~9 行的查询使用:id、:name 这种方式来进行参数绑定。注意：这里使用的列名是属性的名称而不是数据库中列的名称。
- 第 10、11 行也是自定义 JPQL 查询，不同的是传参方式使用?1、?2 这种方式。注意：方法中参数的顺序要与参数声明的顺序一致。
- 如果 BookDao 中的方法涉及修改操作，就需要添加@Modifying 注解并添加事务。

6. 创建 BookService

创建 BookService，代码如下：

```
1   @Service
2   public class BookService {
3       @Autowired
4       BookDao bookDao;
5       public void addBook(Book book) {
6           bookDao.save(book);
7       }
8       public Page<Book> getBookByPage(Pageable pageable) {
9           return bookDao.findAll(pageable);
10      }
11      public List<Book> getBooksByAuthorStartingWith(String author){
12          return bookDao.getBooksByAuthorStartingWith(author);
13      }
14      public List<Book> getBooksByPriceGreaterThan(Float price){
15          return bookDao.getBooksByPriceGreaterThan(price);
16      }
17      public Book getMaxIdBook(){
18          return bookDao.getMaxIdBook();
19      }
20      public List<Book> getBookByIdAndAuthor(String author, Integer id){
21          return bookDao.getBookByIdAndAuthor(author, id);
22      }
23      public List<Book> getBooksByIdAndName(String name, Integer id){
24          return bookDao.getBooksByIdAndName(name, id);
25      }
26  }
```

代码解释:

- 第 6 行使用 save 方法将对象数据保存到数据库,save 方法是由 JpaRepository 接口提供的。
- 第 9 行是一个分页查询,使用 findAll 方法,返回值为 Page<Book>,该对象中包含有分页常用数据,例如总记录数、总页数、每页记录数、当前页记录数等。

7. 创建 BookController

创建 BookController,实现对数据的测试:

```
1   @RestController
2   public class BookController {
3       @Autowired
4       BookService bookService;
5       @GetMapping("/findAll")
6       public void findAll() {
7           PageRequest pageable = PageRequest.of(2, 3);
8           Page<Book> page = bookService.getBookByPage(pageable);
9           System.out.println("总页数:"+page.getTotalPages());
10          System.out.println("总记录数:"+page.getTotalElements());
11          System.out.println("查询结果:"+page.getContent());
12          System.out.println("当前页数:"+(page.getNumber()+1));
13          System.out.println("当前页记录数:"+page.getNumberOfElements());
14          System.out.println("每页记录数:"+page.getSize());
```

```
15          }
16      @GetMapping("/search")
17      public void search() {
18          List<Book> bs1 = bookService.getBookByIdAndAuthor("鲁迅", 7);
19          List<Book> bs2 = bookService.getBooksByAuthorStartingWith("吴");
20          List<Book> bs3 = bookService.getBooksByIdAndName("西", 8);
21          List<Book> bs4 = bookService.getBooksByPriceGreaterThan(30F);
22          Book b = bookService.getMaxIdBook();
23          System.out.println("bs1:"+bs1);
24          System.out.println("bs2:"+bs2);
25          System.out.println("bs3:"+bs3);
26          System.out.println("bs4:"+bs4);
27          System.out.println("b:"+b);
28      }
29      @GetMapping("/save")
30      public void save() {
31          Book book = new Book();
32          book.setAuthor("鲁迅");
33          book.setName("呐喊");
34          book.setPrice(23F);
35          bookService.addBook(book);
36      }
37  }
```

代码解释：

- 在 findAll 接口中，首先通过调用 PageRequest 中的 of 方法构造 PageRequest 对象。of 方法接收两个参数：第一个参数是页数，从 0 开始计；第二个参数是每页显示的条数。
- 在 save 接口中构造一个 Book 对象，直接调用 save 方法保存起来即可。

8. 测试

最后调用相关接口进行测试，数据库中的测试数据如图 5-4 所示。

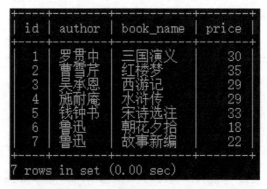

图 5-4

首先调用 http://localhost:8080/findAll 接口，控制台打印日志如图 5-5 所示。

```
总页数:3
总记录数:7
查询结果:[Book{id=7, name='故事新编', author='鲁迅', price=22.0, description='null'}]
当前页数:3
当前页记录数:1
每页记录数:3
```

图 5-5

接下来调用 http://localhost:8080/save 接口，调用后数据库中的数据如图 5-6 所示。

```
+----+--------+-----------+-------+
| id | author | book_name | price |
+----+--------+-----------+-------+
|  1 | 罗贯中  | 三国演义   | 30    |
|  2 | 曹雪芹  | 红楼梦     | 35    |
|  3 | 吴承恩  | 西游记     | 29    |
|  4 | 施耐庵  | 水浒传     | 29    |
|  5 | 钱钟书  | 宋诗选注   | 33    |
|  6 | 鲁迅    | 朝花夕拾   | 18    |
|  7 | 鲁迅    | 故事新编   | 22    |
|  8 | 鲁迅    | 呐喊       | 23    |
+----+--------+-----------+-------+
8 rows in set (0.00 sec)
```

图 5-6

最后调用 http://localhost:8080/search 接口，控制台打印日志如图 5-7 所示。

```
Hibernate: select * from t_book where id=(select max(id) from t_book)
bs1:[Book{id=8, name='呐喊', author='鲁迅', price=23.0, description='null'}]
bs2:[Book{id=3, name='西游记', author='吴承恩', price=29.0, description='null'}]
bs3:[Book{id=3, name='西游记', author='吴承恩', price=29.0, description='null'}]
bs4:[Book{id=2, name='红楼梦', author='曹雪芹', price=35.0, description='null'}, Book{id=5,
 name='宋诗选注', author='钱钟书', price=33.0, description='null'}]
b:Book{id=8, name='呐喊', author='鲁迅', price=23.0, description='null'}
```

图 5-7

5.4 多数据源

所谓多数据源，就是一个 Java EE 项目中采用了不同数据库实例中的多个库，或者同一个数据库实例中多个不同的库。一般来说，采用 MyCat 等分布式数据库中间件是比较好的解决方案，这样可以把数据库读写分离、分库分表、备份等操作交给中间件去做，Java 代码只需要专注于业务即可。不过，这并不意味着无法使用 Java 代码解决类似的问题，在 Spring Framework 中就可以配置多数据源，Spring Boot 继承其衣钵，只不过配置方式有所变化。

5.4.1 JdbcTemplate 多数据源

JdbcTemplate 多数据源的配置是比较简单的,因为一个 JdbcTemplate 对应一个 DataSource,开发者只需要手动提供多个 DataSource,再手动配置 JdbcTemplate 即可。具体步骤如下。

1. 创建数据库

创建两个数据库:chapter05-1 和 chapter05-2。两个库中都创建 book 表,再各预设 1 条数据,创建脚本如下:

```sql
CREATE DATABASE `chapter05-1` DEFAULT CHARACTER SET utf8;
use `chapter05-1`;
CREATE TABLE `book` (
  `id` int(11) NOT NULL AUTO_INCREMENT,
  `name` varchar(128) DEFAULT NULL,
  `author` varchar(128) DEFAULT NULL,
  PRIMARY KEY (`id`)
) ENGINE=InnoDB AUTO_INCREMENT=2 DEFAULT CHARSET=utf8;
insert  into `book`(`id`,`name`,`author`) values (1,'水浒传','施耐庵');

CREATE DATABASE `chapter05-2` DEFAULT CHARACTER SET utf8;
use `chapter05-2`;
CREATE TABLE `book` (
  `id` int(11) NOT NULL AUTO_INCREMENT,
  `name` varchar(128) DEFAULT NULL,
  `author` varchar(128) DEFAULT NULL,
  PRIMARY KEY (`id`)
) ENGINE=InnoDB AUTO_INCREMENT=3 DEFAULT CHARSET=utf8;
insert  into `book`(`id`,`name`,`author`) values (1,'三国演义','罗贯中');
```

执行完数据库脚本后,数据库中的数据如图 5-8 所示。

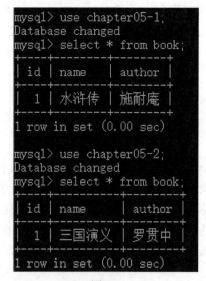

图 5-8

2. 创建项目

创建 Spring Boot Web 项目，添加如下依赖：

```
1   <dependency>
2       <groupId>org.springframework.boot</groupId>
3       <artifactId>spring-boot-starter-jdbc</artifactId>
4   </dependency>
5   <dependency>
6       <groupId>org.springframework.boot</groupId>
7       <artifactId>spring-boot-starter-web</artifactId>
8   </dependency>
9   <dependency>
10      <groupId>com.alibaba</groupId>
11      <artifactId>druid-spring-boot-starter</artifactId>
12      <version>1.1.10</version>
13  </dependency>
14  <dependency>
15      <groupId>mysql</groupId>
16      <artifactId>mysql-connector-java</artifactId>
17      <scope>runtime</scope>
18  </dependency>
```

注意这里添加的数据库连接池依赖是 druid-spring-boot-starter。druid-spring-boot-starter 可以帮助开发者在 Spring Boot 项目中轻松集成 Druid 数据库连接池和监控。

3. 配置数据库连接

在 application.properties 中配置数据库连接信息，代码如下：

```
1   # 数据源1
2   spring.datasource.one.type=com.alibaba.druid.pool.DruidDataSource
3   spring.datasource.one.username=root
4   spring.datasource.one.password=123
5   spring.datasource.one.url=jdbc:mysql:///chapter05-1
6   # 数据源2
7   spring.datasource.two.type=com.alibaba.druid.pool.DruidDataSource
8   spring.datasource.two.username=root
9   spring.datasource.two.password=123
10  spring.datasource.two.url=jdbc:mysql:///chapter05-2
```

配置两个数据源，区别主要是数据库不同，其他都是一样的。

4. 配置数据源

创建 DataSourceConfig 配置数据源，根据 application.properties 中的配置生成两个数据源：

```
1   @Configuration
2   public class DataSourceConfig {
3       @Bean
4       @ConfigurationProperties("spring.datasource.one")
5       DataSource dsOne() {
6           return DruidDataSourceBuilder.create().build();
```

```
7         }
8         @Bean
9         @ConfigurationProperties("spring.datasource.two")
10        DataSource dsTwo() {
11            return DruidDataSourceBuilder.create().build();
12        }
13    }
```

代码解释：

- DataSourceConfig 中提供了两个数据源：dsOne 和 dsTwo，默认方法名即实例名。
- @ConfigurationProperties 注解表示使用不同前缀的配置文件来创建不同的 DataSource 实例。

5. 配置 JdbcTemplate

在 5.1 节中，读者已经了解到只要引入了 spring-jdbc 依赖，那么开发者没有提供 JdbcTemplate 实例时，Spring Boot 默认会提供一个 JdbcTemplate 实例。现在配置多数据源时，由开发者自己提供 JdbcTemplate 实例，代码如下：

```
1   @Configuration
2   public class JdbcTemplateConfig {
3       @Bean
4       JdbcTemplate jdbcTemplateOne(@Qualifier("dsOne")DataSource dataSource) {
5           return new JdbcTemplate(dataSource);
6       }
7       @Bean
8       JdbcTemplate jdbcTemplateTwo(@Qualifier("dsTwo")DataSource dataSource) {
9           return new JdbcTemplate(dataSource);
10      }
11  }
```

代码解释：

- JdbcTemplateConfig 中提供两个 JdbcTemplate 实例。每个 JdbcTemplate 实例都需要提供 DataSource，由于 Spring 容器中有两个 DataSource 实例，因此需要通过方法名查找。@Qualifier 注解表示查找不同名称的 DataSource 实例注入进来。

6. 创建 BookController

创建实体类 Book 和 BookController 进行测试：

```
1   public class Book {
2       private Integer id;
3       private String name;
4       private String author;
5       //省略getter/setter
6   }
7   @RestController
8   public class BookController {
9       @Resource(name = "jdbcTemplateOne")
10      JdbcTemplate jdbcTemplateOne;
11      @Autowired
```

```
12        @Qualifier("jdbcTemplateTwo")
13        JdbcTemplate jdbcTemplateTwo;
14        @GetMapping("/test1")
15        public void test1() {
16            List<Book> books1 = jdbcTemplateOne.query("select * from book",
17                    new BeanPropertyRowMapper<>(Book.class));
18            List<Book> books2 = jdbcTemplateTwo.query("select * from book",
19                    new BeanPropertyRowMapper<>(Book.class));
20            System.out.println("books1:"+books1);
21            System.out.println("books2:"+books2);
22        }
23  }
```

简单起见，这里没有添加 Service 层，而是直接将 JdbcTemplate 注入到了 Controller 中。在 Controller 中注入两个不同的 JdbcTemplate 有两种方式：一种是使用@Resource 注解，并指明 name 属性，即按 name 进行装配，此时会根据实例名查找相应的实例注入；另一种是使用@Autowired 注解结合@Qualifier 注解，效果等同于使用@Resource 注解。

7．测试

最后，在浏览器地址栏输入"http://localhost:8080/test1"，控制台打印日志如图 5-9 所示。JdbcTemplate 多数据源配置成功。

```
2018-07-16 22:40:07.194  INFO 120 --- [nio-8080-exec-2]
books1:[Book{id=1, name='水浒传', author='施耐庵'}]
books2:[Book{id=1, name='三国演义', author='罗贯中'}]
```

图 5-9

5.4.2　MyBatis 多数据源

JdbcTemplate 可以配置多数据源，MyBatis 也可以配置，但是步骤要稍微复杂一些。

1．准备工作

本案例使用的数据库与 5.4.1 小节一致，这里不再赘述。创建的项目也和 5.4.1 小节一致，只不过将 spring-boot-starter-jdbc 依赖换成如下 MyBatis 依赖：

```
1  <dependency>
2      <groupId>org.mybatis.spring.boot</groupId>
3      <artifactId>mybatis-spring-boot-starter</artifactId>
4      <version>1.3.2</version>
5  </dependency>
```

另外，application.properties 中两个数据源的配置、DataSourceConfig 以及 Book 实体类也和 5.4.1 小节一致，这里不赘述。同时，为了使 Mapper 映射文件不被过滤掉，pom.xml 中的配置与 5.2 节中的第 6 步配置一致。

2. 创建 MyBatis 配置

配置 MyBatis，主要提供 SqlSessionFactory 实例和 SqlSessionTemplate 实例，代码如下：

```
@Configuration
@MapperScan(value = "org.sang.mapper1", sqlSessionFactoryRef =
"sqlSessionFactoryBean1")
public class MyBatisConfigOne {
    @Autowired
    @Qualifier("dsOne")
    DataSource dsOne;
    @Bean
    SqlSessionFactory sqlSessionFactoryBean1() throws Exception {
        SqlSessionFactoryBean factoryBean = new SqlSessionFactoryBean();
        factoryBean.setDataSource(dsOne);
        return factoryBean.getObject();
    }
    @Bean
    SqlSessionTemplate sqlSessionTemplate1() throws Exception {
        return new SqlSessionTemplate(sqlSessionFactoryBean1());
    }
}
```

代码解释：

- 在@MapperScan 注解中指定 Mapper 接口所在的位置，同时指定 SqlSessionFactory 的实例名，则该位置下的 Mapper 将使用 SqlSessionFactory 实例。
- 提供 SqlSessionFactory 的实例，直接创建出来，同时将 DataSource 的实例设置给 SqlSessionFactory，这里创建的 SqlSessionFactory 实例也就是@MapperScan 注解中 sqlSessionFactoryRef 参数指定的实例。
- 提供一个 SqlSessionTemplate 实例。这是一个线程安全类，主要用来管理 MyBatis 中的 SqlSession 操作。

当 MyBatisConfigOne 创建成功后，参考 MyBatisConfigOne 创建 MyBatisConfigTwo，代码如下：

```
@Configuration
@MapperScan(value = "org.sang.mapper2", sqlSessionFactoryRef =
"sqlSessionFactoryBean2")
public class MyBatisConfigTwo {
    @Autowired
    @Qualifier("dsTwo")
    DataSource dsTwo;
    @Bean
    SqlSessionFactory sqlSessionFactoryBean2() throws Exception {
        SqlSessionFactoryBean factoryBean = new SqlSessionFactoryBean();
        factoryBean.setDataSource(dsTwo);
        return factoryBean.getObject();
    }
    @Bean
```

```
15      SqlSessionTemplate sqlSessionTemplate2() throws Exception {
16          return new SqlSessionTemplate(sqlSessionFactoryBean2());
17      }
18  }
```

3. 创建 Mapper

分别在 org.sang.mapper1 和 org.sang.mapper2 包下创建两个不同的 Mapper 以及相应的 Mapper 映射文件，代码如下。

org.sang.mapper1 中：

```
1   public interface BookMapper {
2       List<Book> getAllBooks();
3   }
4   <?xml version="1.0" encoding="UTF-8" ?>
5   <!DOCTYPE mapper
6           PUBLIC "-//mybatis.org//DTD Mapper 3.0//EN"
7   "http://mybatis.org/dtd/mybatis-3-mapper.dtd">
8   <mapper namespace="org.sang.mapper1.BookMapper">
9   <select id="getAllBooks" resultType="org.sang.model.Book">
10      select * from book;
11  </select>
12  </mapper>
```

org.sang.mapper2 中：

```
1   public interface BookMapper2 {
2       List<Book> getAllBooks();
3   }
4   <?xml version="1.0" encoding="UTF-8" ?>
5   <!DOCTYPE mapper
6           PUBLIC "-//mybatis.org//DTD Mapper 3.0//EN"
7   "http://mybatis.org/dtd/mybatis-3-mapper.dtd">
8   <mapper namespace="org.sang.mapper2.BookMapper2">
9   <select id="getAllBooks" resultType="org.sang.model.Book">
10      select * from book;
11  </select>
12  </mapper>
```

这两个不同的 Mapper 将操作不同的数据源。

4. 创建 Controller

简便起见，这里直接将 Mapper 注入 Controller 中，代码如下：

```
1   @RestController
2   public class BookController {
3       @Autowired
4       BookMapper bookMapper;
5       @Autowired
6       BookMapper2 bookMapper2;
7       @GetMapping("/test1")
8       public void test1() {
```

```
 9          List<Book> books1 = bookMapper.getAllBooks();
10          List<Book> books2 = bookMapper2.getAllBooks();
11          System.out.println("books1:"+books1);
12          System.out.println("books2:"+books2);
13      }
14  }
```

在 Controller 中注入两个不同的 Mapper，然后调用两个 Mapper 中的查询方法。

5. 测试

最后，在浏览器中输入 http://localhost:8080/test1，即可看到控制台打印了不同数据库中的数据，如图 5-10 所示。

```
Tue Jul 17 08:10:18 CST 2018 WARN: Establishing SSL con
books1:[Book{id=1, name='水浒传', author='施耐庵'}]
books2:[Book{id=1, name='三国演义', author='罗贯中'}]
```

图 5-10

5.4.3 JPA 多数据源

JPA 和 MyBatis 配置多数据源类似，不同的是，JPA 配置时主要提供不同的 LocalContainerEntityManagerFactoryBean 以及事务管理器，具体配置步骤如下。

1. 准备工作

本案例使用的数据库与 5.4.1 小节一致，这里不再赘述。创建的项目也和 5.4.1 小节一致，只不过将 spring-boot-starter-jdbc 依赖换成如下 Spring Data JPA 依赖：

```
1  <dependency>
2      <groupId>org.springframework.boot</groupId>
3      <artifactId>spring-boot-starter-data-jpa</artifactId>
4  </dependency>
```

application.properties 中两个数据源在原有配置的基础上再添加如下 JPA 相关的配置：

```
1  spring.jpa.properties.hibernate.dialect=org.hibernate.dialect.MySQL57InnoDBDialect
2  
3  spring.jpa.properties.database=mysql
4  spring.jpa.properties.hibernate.hbm2ddl.auto=update
   spring.jpa.properties.show-sql= true
```

> **注意**
>
> 这里的配置与配置单独的 JPA 有区别，因为在后文的配置中要从 JpaProperties 中的 getProperties 方法中获取所有 JPA 相关的配置，因此这里的属性前缀都是 spring.jpa.properties。

DataSourceConfig 也和 5.4.1 小节基本一致，不同的是，这里多了一个@Primary 注解，代码如下：

```
1   @Configuration
2   public class DataSourceConfig {
3       @Bean
4       @ConfigurationProperties("spring.datasource.one")
5       @Primary
6       DataSource dsOne() {
7           return DruidDataSourceBuilder.create().build();
8       }
9       @Bean
10      @ConfigurationProperties("spring.datasource.two")
11      DataSource dsTwo() {
12          return DruidDataSourceBuilder.create().build();
13      }
14  }
```

2. 创建实体类

在 org.sang.model 包下创建实体类 User，代码如下：

```
1   @Entity(name = "t_user")
2   public class User {
3       @Id
4       @GeneratedValue(strategy = GenerationType.IDENTITY)
5       private Integer id;
6       private String name;
7       private String gender;
8       private Integer age;
9       //省略 getter/setter
10  }
```

根据实体类在数据库中创建 t_user 表，表中的 id 字段自增长。

3. 创建 JPA 配置

接下来是核心配置，根据两个配置好的数据源创建两个不同的 JPA 配置，代码如下：

```
1   @Configuration
2   @EnableTransactionManagement
3   @EnableJpaRepositories(basePackages = "org.sang.dao1",
4           entityManagerFactoryRef = "entityManagerFactoryBeanOne",
5           transactionManagerRef = "platformTransactionManagerOne")
6   public class JpaConfigOne {
7       @Resource(name = "dsOne")
8       DataSource dsOne;
9       @Autowired
10      JpaProperties jpaProperties;
11      @Bean
12      @Primary
13      LocalContainerEntityManagerFactoryBean entityManagerFactoryBeanOne(
```

```
14              EntityManagerFactoryBuilder builder) {
15          return builder.dataSource(dsOne)
16                  .properties(jpaProperties.getProperties())
17                  .packages("org.sang.model")
18                  .persistenceUnit("pu1")
19                  .build();
20      }
21      @Bean
22      PlatformTransactionManager platformTransactionManagerOne(
23              EntityManagerFactoryBuilder builder) {
24          LocalContainerEntityManagerFactoryBean factoryOne =
25  entityManagerFactoryBeanOne(builder);
26          return new JpaTransactionManager(factoryOne.getObject());
27      }
28  }
```

代码解释：

- 使用 @EnableJpaRepositories 注解来进行 JPA 的配置，该注解中主要配置三个属性：basePackages、entityManagerFactoryRef 以及 transactionManagerRef。其中，basePackages 用来指定 Repository 所在的位置，entityManagerFactoryRef 用来指定实体类管理工厂 Bean 的名称，transactionManagerRef 则用来指定事务管理器的引用名称，这里的引用名称就是 JpaConfigOne 类中注册的 Bean 的名称（默认的 Bean 名称为方法名）。
- 第 13~20 行创建 LocalContainerEntityManagerFactoryBean，该 Bean 将用来提供 EntityManager 实例，在该类的创建过程中，首先配置数据源，然后设置 JPA 相关配置（JpaProperties 由系统自动加载），再设置实体类所在的位置，最后配置持久化单元名，若项目中只有一个 EntityManagerFactory，则 persistenceUnit 可以省略掉，若有多个，则必须明确指定持久化单元名。
- 由于项目中会提供两个 LocalContainerEntityManagerFactoryBean 实例，第 12 行的注解 @Primary 表示当存在多个 LocalContainerEntityManagerFactoryBean 实例时，该实例将被优先使用。
- 第 21~27 行表示创建一个事务管理器。JpaTransactionManager 提供对单个 EntityManagerFactory 的事务支持，专门用于解决 JPA 中的事务管理。

这是第一个 JPA 配置，第二个与之类似，代码如下：

```
1   @Configuration
2   @EnableTransactionManagement
3   @EnableJpaRepositories(basePackages = "org.sang.dao2",
4   entityManagerFactoryRef = "entityManagerFactoryBeanTwo",
5   transactionManagerRef = "platformTransactionManagerTwo")
6   public class JpaConfigTwo {
7       @Resource(name = "dsTwo")
8       DataSource dsTwo;
9       @Autowired
10      JpaProperties jpaProperties;
```

```
11      @Bean
12      LocalContainerEntityManagerFactoryBean entityManagerFactoryBeanTwo(
13              EntityManagerFactoryBuilder builder) {
14          return builder.dataSource(dsTwo)
15                  .properties(jpaProperties.getProperties())
16                  .packages("org.sang.model")
17                  .persistenceUnit("pu2")
18                  .build();
19      }
20      @Bean
21      PlatformTransactionManager platformTransactionManagerTwo(
22              EntityManagerFactoryBuilder builder) {
23          LocalContainerEntityManagerFactoryBean factoryTwo =
24  entityManagerFactoryBeanTwo(builder);
25          return new JpaTransactionManager(factoryTwo.getObject());
26      }
27  }
```

JpaConfigTwo 的配置与 JpaConfigOne 类似，注意 LocalContainerEntityManagerFactoryBean 实例不需要添加@Primary 注解。

4. 创建 Repository

根据第 3 步的配置，分别在 org.sang.dao1 和 org.sang.dao2 包下创建两个 Repository。

UserDao 如下：

```
1   package org.sang.dao1;
2   import org.sang.model.User;
3   import org.springframework.data.jpa.repository.JpaRepository;
4   public interface UserDao extends JpaRepository<User,Integer>{
5   }
```

UserDao2 如下：

```
1   package org.sang.dao2;
2   import org.sang.model.User;
3   import org.springframework.data.jpa.repository.JpaRepository;
4   public interface UserDao2 extends JpaRepository<User,Integer>{
5   }
```

UserDao 和 UserDao2 将操作不同的数据源。

5. 创建 Controller

简便起见，这里省略掉 Service 层，将 UserDao 直接注入 Controller 中，代码如下：

```
1   @RestController
2   public class UserController {
3       @Autowired
4       UserDao userDao;
5       @Autowired
6       UserDao2 userDao2;
7       @GetMapping("/test1")
```

```
 8      public void test1() {
 9          User u1 = new User();
10          u1.setAge(55);
11          u1.setName("鲁迅");
12          u1.setGender("男");
13          userDao.save(u1);
14          User u2 = new User();
15          u2.setAge(80);
16          u2.setName("泰戈尔");
17          u2.setGender("男");
18          userDao2.save(u2);
19      }
20  }
```

6. 测试

在浏览器中输入"http://localhost:8080/test1",然后查看数据库,即可看到数据库中的表和数据都已经存在了,如图 5-11 所示。

图 5-11

5.5 小　　结

本章主要和读者分享了 Spring Boot 整合持久层技术,包括 JdbcTemplate、MyBatis 以及 Spring Data JPA。其中,JdbcTemplate 使用得并不是很广泛;MyBatis 灵活性较好,方便开发者进行 SQL 优化;Spring Data JPA 使用方便,特别是快速实现一个 RESTful 风格的应用(将在第 7 章向读者介绍)。

第 6 章

Spring Boot 整合 NoSQL

本章概要

- 整合 Redis
- 整合 MongoDB
- Session 共享

NoSQL 是指非关系型数据库，非关系型数据库和关系型数据库两者存在许多显著的不同点，其中最重要的是 NoSQL 不使用 SQL 作为查询语言。其数据存储可以不需要固定的表格模式，一般都有水平可扩展性的特征。NoSQL 主要有如下几种不同的分类：

- Key/Value 键值存储。这种数据存储通常都是无数据结构的，一般被当作字符串或者二进制数据，但是数据加载速度快，典型的使用场景是处理高并发或者用于日志系统等，这一类的数据库有 Redis、Tokyo Cabinet 等。
- 列存储数据库。列存储数据库功能相对局限，但是查找速度快，容易进行分布式扩展，一般用于分布式文件系统中，这一类的数据库有 HBase、Cassandra 等。
- 文档型数据库。和 Key/Value 键值存储类似，文档型数据库也没有严格的数据格式，这既是缺点也是优势，因为不需要预先创建表结构，数据格式更加灵活，一般可用在 Web 应用中，这一类数据库有 MongoDB、CouchDB 等。
- 图形数据库。图形数据库专注于构建关系图谱，例如社交网络，推荐系统等，这一类的数据库有 Neo4J、DEX 等。

NoSQL 种类繁多，Spring Boot 对大多数 NoSQL 都提供了配置支持，本书主要向读者介绍常见的两个：Redis 和 MongoDB。

6.1 整合 Redis

6.1.1 Redis 简介

Redis 是一个使用 C 编写的基于内存的 NoSQL 数据库，它是目前最流行的键值对存储数据库。Redis 由一个 Key、Value 映射的字典构成，与其他 NoSQL 不同，Redis 中 Value 的类型不局限于字符串，还支持列表、集合、有序集合、散列等。Redis 不仅可以当作缓存使用，也可以配置数据持久化后当作 NoSQL 数据库使用，目前支持两种持久化方式：快照持久化和 AOF 持久化。另一方面，Redis 也可以搭建集群或者主从复制结构，在高并发环境下具有高可用性。

6.1.2 Redis 安装

Redis 版本使用写作本书时的最新版 4.0.10，安装环境选择 CentOS 7。安装步骤如下。

1. 下载 Redis

首先执行如下命令下载 Redis：

```
wget http://download.redis.io/releases/redis-4.0.10.tar.gz
```

若提示未找到命令，则先执行如下命令安装 wget，再下载 Redis：

```
yum install wget
```

2. 安装 Redis

首先解压下载的文件，然后进入解压目录中进行编译，执行如下 4 条命令：

```
tar -zxvf redis-4.0.10.tar.gz
cd redis-4.0.10
make MALLOC=libc
make install
```

若在执行 make MALLOC=libc 命令时提示"gcc：未找到命令"，则先安装 gcc，命令如下：

```
yum install gcc
```

安装成功后再进行编译安装。

3. 配置 Redis

Redis 安装成功后，接下来进行配置，打开 Redis 解压目录下的 redis.conf 文件，主要修改如下几个地方：

```
daemonize yes
#bind 127.0.0.1
requirepass 123@456
protected-mode no
```

配置解释：
- 第 1 行配置表示允许 Redis 在后台启动。
- 第 2 行配置表示允许连接该 Redis 实例的地址，默认情况下只允许本地连接，将默认配置注释掉，外网就可以连接 Redis 了。
- 第 3 行配置表示登录该 Redis 实例所需的密码。
- 由于有了第 3 行配置的密码登录，因此第 4 行就可以关闭保护模式了。

4. 配置 CentOS

为了能够远程连接上 Redis，还需要关闭 CentOS 防火墙，执行如下命令：

```
1  systemctl stop firewalld.service
2  systemctl disable firewalld.service
```

其中，第 1 行表示关闭防火墙，第 2 行表示禁止防火墙开机启动。

5. Redis 启动与关闭

最后，执行如下命令启动 Redis：

```
1  redis-server redis.conf
```

Redis 启动成功后，再执行如下命令进入 Redis 控制台，其中 -a 表示 Redis 登录密码：

```
1  redis-cli -a 123@456
```

进入控制台后执行 ping 命令，如果能看到 PONG，表示 Redis 安装成功，如图 6-1 所示。

如果想关闭 Redis 实例，可以在控制台执行 SHUTDOWN，然后使用 exit 退出（如图 6-2 所示），或者直接在命令行执行如下命令：

```
1  redis-cli -p 6379 -a 123@456 shutdown
```

其中，-p 表示要关闭的 Redis 实例的端口号，-a 表示 Redis 登录密码。

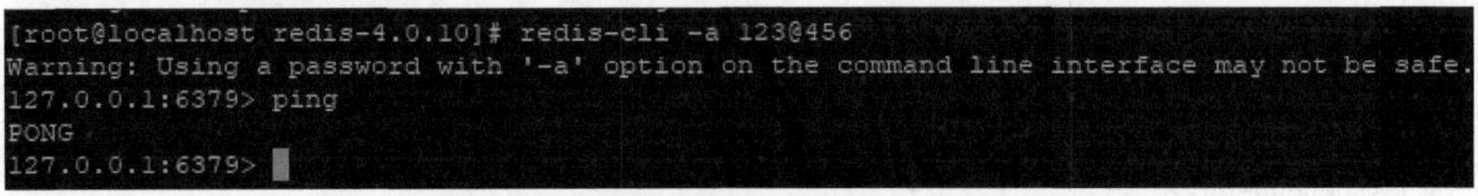

图 6-1

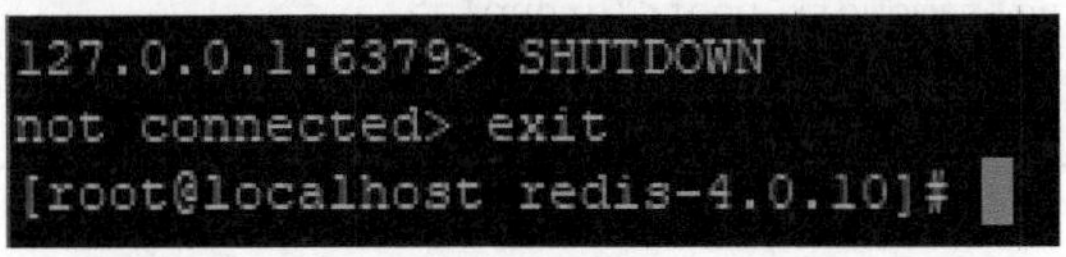

图 6-2

至此，单机版 Redis 就安装并启动成功了。

6.1.3 整合 Spring Boot

Redis 的 Java 客户端有很多，例如 Jedis、JRedis、Spring Data Redis 等，Spring Boot 借助于 Spring Data Redis 为 Redis 提供了开箱即用自动化配置，开发者只需要添加相关依赖并配置 Redis 连接信息即可，具体整合步骤如下。

1. 创建 Spring Boot 项目

首先创建 Spring Boot Web 项目，添加如下依赖：

```
1  <dependency>
2      <groupId>org.springframework.boot</groupId>
3      <artifactId>spring-boot-starter-data-redis</artifactId>
4  </dependency>
5  <dependency>
6      <groupId>org.springframework.boot</groupId>
7      <artifactId>spring-boot-starter-web</artifactId>
8  </dependency>
```

默认情况下，spring-boot-starter-data-redis 使用的 Redis 工具是 Lettuce，考虑到有的开发者习惯使用 Jedis，因此可以从 spring-boot-starter-data-redis 中排除 Lettuce 并引入 Jedis，修改为如下依赖：

```
1   <dependency>
2       <groupId>org.springframework.boot</groupId>
3       <artifactId>spring-boot-starter-data-redis</artifactId>
4       <exclusions>
5           <exclusion>
6               <groupId>io.lettuce</groupId>
7               <artifactId>lettuce-core</artifactId>
8           </exclusion>
9       </exclusions>
10  </dependency>
11  <dependency>
12      <groupId>redis.clients</groupId>
13      <artifactId>jedis</artifactId>
14  </dependency>
15  <dependency>
16      <groupId>org.springframework.boot</groupId>
17      <artifactId>spring-boot-starter-web</artifactId>
18  </dependency>
```

2. 配置 Redis

接下来在 application.properties 中配置 Redis 连接信息，代码如下：

```
1  spring.redis.database=0
2  spring.redis.host=192.168.248.144
3  spring.redis.port=6379
4  spring.redis.password=123@456
5  spring.redis.jedis.pool.max-active=8
```

```
6    spring.redis.jedis.pool.max-idle=8
7    spring.redis.jedis.pool.max-wait=-1ms
8    spring.redis.jedis.pool.min-idle=0
```

配置解释：

- 第 1~4 行是基本连接信息配置，第 5~8 行是连接池信息配置。
- 第 1 行配置表示使用的 Redis 库的编号，Redis 中提供了 16 个 database，编号为 0~15。
- 第 2 行配置表示 Redis 实例的地址。
- 第 3 行配置表示 Redis 端口号，默认是 6379。
- 第 4 行配置表示 Redis 登录密码。
- 第 5 行配置表示 Redis 连接池的最大连接数。
- 第 6 行配置表示 Redis 连接池中的最大空闲连接数。
- 第 7 行配置表示连接池的最大阻塞等待时间，默认为-1，表示没有限制。
- 第 8 行配置表示连接池最小空闲连接数。
- 如果项目使用了 Lettuce，则只需将第 5~8 行配置中的 jedis 修改为 lettuce 即可。

在 Spring Boot 的自动配置类中提供了 RedisAutoConfiguration 进行 Redis 的配置，部分源码如下：

```
1    @Configuration
2    @ConditionalOnClass(RedisOperations.class)
3    @EnableConfigurationProperties(RedisProperties.class)
4    @Import({LettuceConnectionConfiguration.class,JedisConnectionConfiguration.class})
5    public class RedisAutoConfiguration {
6       @Bean
7       @ConditionalOnMissingBean(name = "redisTemplate")
8       public RedisTemplate<Object, Object> redisTemplate(
9       …
10      …
11   return template;
12   }
13   @Bean
14   @ConditionalOnMissingBean
15   public StringRedisTemplate stringRedisTemplate(
16      …
17      …
18   return template;
19   }
20   }
```

由这一段源码可以看到，application.properties 中配置的信息将被注入 RedisProperties 中，如果开发者自己没有提供 RedisTemplate 或者 StringRedisTemplate 实例，则 Spring Boot 默认会提供这两个实例，RedisTemplate 和 StringRedisTemplate 实例则提供了 Redis 的基本操作方法。

3. 创建实体类

创建一个 Book 类，代码如下：

```
public class Book implements Serializable {
    private Integer id;
    private String name;
    private String author;
    //省略 getter/setter
}
```

4. 创建 Controller

创建 BookController 进行测试：

```
@RestController
public class BookController {
    @Autowired
    RedisTemplate redisTemplate;
    @Autowired
    StringRedisTemplate stringRedisTemplate;
    @GetMapping("/test1")
    public void test1() {
        ValueOperations<String, String> ops1 = stringRedisTemplate.opsForValue();
        ops1.set("name", "三国演义");
        String name = ops1.get("name");
        System.out.println(name);
        ValueOperations ops2 = redisTemplate.opsForValue();
        Book b1 = new Book();
        b1.setId(1);
        b1.setName("红楼梦");
        b1.setAuthor("曹雪芹");
        ops2.set("b1", b1);
        Book book = (Book) ops2.get("b1");
        System.out.println(book);
    }
}
```

代码解释：

- StringRedisTemplate 是 RedisTemplate 的子类，StringRedisTemplate 中的 key 和 value 都是字符串，采用的序列化方案是 StringRedisSerializer，而 RedisTemplate 则可以用来操作对象，RedisTemplate 采用的序列化方案是 JdkSerializationRedisSerializer。无论是 StringRedisTemplate 还是 RedisTemplate，操作 Redis 的方法都是一致的。
- StringRedisTemplate 和 RedisTemplate 都是通过 opsForValue、opsForZSet 或者 opsForSet 等方法首先获取一个操作对象，再使用该操作对象完成数据的读写。
- 第 10 行向 Redis 中存储一条记录，第 11 行将之读取出来。第 18 行向 Redis 中存储一个对象，第 19 行将之读取出来。

5. 测试

在浏览器中输入 http://localhost:8080/test1，可看到控制打印日志如图 6-3 所示。

```
2018-07-18 16:17:39.031  INFO 10768 --- [ni
三国演义
Book{id=1, name='红楼梦', author='曹雪芹'}
```

图 6-3

6.1.4 Redis 集群整合 Spring Boot

前文向读者介绍了单个 Redis 实例整合 Spring Boot，在实际项目中，开发者为了提高 Redis 的扩展性，往往需要搭建 Redis 集群，这样就会涉及 Redis 集群整合 Spring Boot，接下来看看这个问题。

1. 搭建 Redis 集群

（1）集群原理

在 Redis 集群中，所有的 Redis 节点彼此互联，节点内部使用二进制协议优化传输速度和带宽。当一个节点挂掉后，集群中超过半数的节点检测失效时才认为该节点已失效。不同于 Tomcat 集群需要使用反向代理服务器，Redis 集群中的任意节点都可以直接和 Java 客户端连接。Redis 集群上的数据分配则是采用哈希槽（HASH SLOT），Redis 集群中内置了 16384 个哈希槽，当有数据需要存储时，Redis 会首先使用 CRC16 算法对 key 进行计算，将计算获得的结果对 16384 取余，这样每一个 key 都会对应一个取值在 0~16383 之间的哈希槽，Redis 则根据这个余数将该条数据存储到对应的 Redis 节点上，开发者可根据每个 Redis 实例的性能来调整每个 Redis 实例上哈希槽的分布范围。

（2）集群规划

本案例在同一台服务器上用不同的端口表示不同的 Redis 服务器（伪分布式集群）。
主节点：192.168.248.144:8001，192.168.248.144:8002，192.168.248.144:8003。
从节点：192.168.248.144:8004，192.168.248.144:8005，192.168.248.144:8006。

（3）集群配置

Redis 集群管理工具 redis-trib.rb 依赖 Ruby 环境，首先需要安装 Ruby 环境，由于 CentOS 7 yum 库中默认的 Ruby 版本较低，因此建议采用如下步骤进行安装。

首先安装 RVM，RVM 是一个命令行工具，可以提供一个便捷的多版本 Ruby 环境的管理和切换，安装命令如下：

```
1  gpg2 --keyserver hkp://keys.gnupg.net --recv-keys D39DC0E3
2  curl -L get.rvm.io | bash -s stable
3  source /usr/local/rvm/scripts/rvm
```

最后一条命令表示安装完后使 RVM 名生效，RVM 安装成功后，查看 RVM 中有哪些 Ruby：

```
1  rvm list known
```

查看结果如图 6-4 所示。

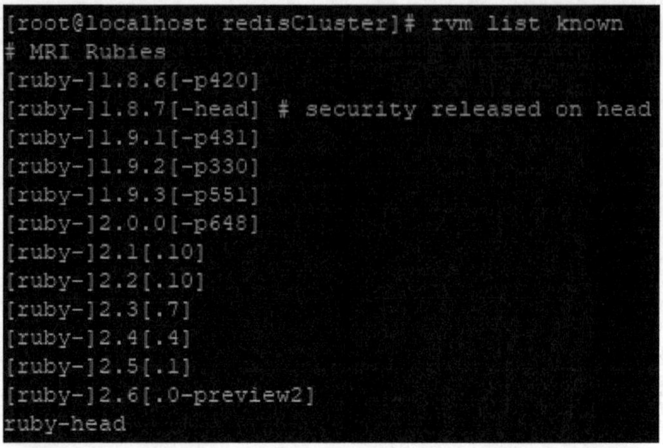

图 6-4

选择最新的稳定版进行安装，命令如下：

```
rvm install 2.5.1
```

最后安装 Redis 依赖，命令如下：

```
gem install redis
```

接下来创建 redisCluster 文件夹，将 6.1.2 节中下载的 Redis 压缩文件复制到 redisCluster 文件夹中之后编译安装，操作命令如下：

```
mkdir redisCluster
cp -f ./redis-4.0.10.tar.gz ./redisCluster/
cd redisCluster
tar -zxvf redis-4.0.10.tar.gz
cd redis-4.0.10
make MALLOC=libc
make install
```

安装成功后，将 redis-4.0.10/src 目录下的 redis-trib.rb 文件复制到 redisCluster 目录下，命令如下：

```
cp -f ./redis-4.0.10/src/redis-trib.rb ./
```

然后在 redisCluster 目录下创建 6 个文件夹，分别命名为 8001、8002、8003、8004、8005、8006，再将 redis-4.0.10 目录下的 redis.conf 文件分别往这 6 个目录中复制一份，然后对每个目录中的 redis.conf 文件进行修改，以 8001 目录下的 redis.conf 文件为例，主要修改如下配置：

```
port 8001
#bind 127.0.0.1
cluster-enabled yes
cluster-config-file nodes-8001.conf
protected no
daemonize yes
requirepass 123@456
masterauth 123@456
```

这里的配置在 6.1.2 小节的单机版安装配置的基础上增加了几条，其中端口修改为 8001，cluster-enabled 表示开启集群，cluster-config-file 表示集群节点的配置文件，由于每个节点都开启了密码认证，因此又增加了 masterauth 配置，使得从机可以登录到主机上。按照这里的配置，对 8002~8006 目录中的 redis.conf 文件依次进行修改，注意修改时每个文件的 port 和 cluster-config-file 不一样。全部修改完成后，进入 redis-4.0.10 目录下，分别启动 6 个 Redis 实例，相关命令如下：

```
1  redis-server ../8001/redis.conf
2  redis-server ../8002/redis.conf
3  redis-server ../8003/redis.conf
4  redis-server ../8004/redis.conf
5  redis-server ../8005/redis.conf
6  redis-server ../8006/redis.conf
```

当 6 个 Redis 实例都启动成功后，回到 redisCluster 目录下，首先对 redis-trib.rb 文件进行修改，由于配置了密码登录，而该命令在执行时默认没有密码，因此将登录不上各个 Redis 实例，此时用 vi 编辑器打开 redis-trib.rb 文件，搜索到如下一行：

```
1  @r = Redis.new(:host => @info[:host], :port => @info[:port], :timeout => 60)
```

修改这一行，添加密码参数：

```
1  @r = Redis.new(:host => @info[:host], :port => @info[:port], :timeout =>
2  60,:password=>"123@456")
```

123@456 就是各个 Redis 实例的登录密码。

这些配置都完成后，接下来就可以创建 Redis 集群了。

（4）创建集群

执行如下命令创建 Redis 集群：

```
1  ./redis-trib.rb create --replicas 1 192.168.248.144:8001 192.168.248.144:8002
   192.168.248.144:8003 192.168.248.144:8004 192.168.248.144:8005 192.168.248.144:8006
```

其中，replicas 表示每个主节点的 slave 数量。在集群的创建过程中会分配主机和从机，每个集群在创建过程中都将分配到一个唯一的 id 并分配到一段 slot。

当集群创建成功后，进入 redis-4.0.10 目录中，登录任意 Redis 实例，命令如下：

```
1  redis-cli -p 8001 -a 123@456 -c
```

-p 表示要登录的集群的端口，-a 表示要登录的集群的密码，-c 则表示以集群的方式登录。登录成功后，通过 cluster info 命令可以查询集群状态信息（如图 6-5 所示），通过 cluster nodes 命令可以查询集群节点信息（如图 6-6 所示），在集群节点信息中，可以看到每一个节点的 id，该节点是 slave 还是 master，如果是 slave，那么它的 master 的 id 是什么，如果是 master，那么每一个 master 的 slot 范围是多少，这些信息都会显示出来。

（5）添加主节点

当集群创建成功后，随着业务的增长，有可能需要添加主节点，添加主节点需要先构建主节点实例，将 redisCluster 目录下的 8001 目录再复制一份，名为 8007，根据第 3 步的集群配置修改 8007 目录下的 redis.conf 文件，修改完成后，在 redis-4.0.10 目录下运行如下命令启动该节点：

```
127.0.0.1:8001> cluster info
cluster_state:ok
cluster_slots_assigned:16384
cluster_slots_ok:16384
cluster_slots_pfail:0
cluster_slots_fail:0
cluster_known_nodes:6
cluster_size:3
cluster_current_epoch:6
cluster_my_epoch:1
cluster_stats_messages_ping_sent:669
cluster_stats_messages_pong_sent:736
cluster_stats_messages_sent:1405
cluster_stats_messages_ping_received:731
cluster_stats_messages_pong_received:669
cluster_stats_messages_meet_received:5
cluster_stats_messages_received:1405
```

图 6-5

```
127.0.0.1:8001> cluster nodes
adc40fd1dd0007fe4b4f7aff47dc52ea8dbe1b7b 192.168.248.144:8006@18006 slave 619297db9a97533c4a5d8d1b57df535e2de5db9c 0 1531912346250 6 connected
7a57fcf36ecc4c0fbaa30efd5545928abd3be81d 192.168.248.144:8001@18001 myself,master - 0 1531912346000 1 connected 0-5460
75ec7962be8bc1d4b207e816458ba2c268d3329a 192.168.248.144:8004@18004 slave 6a84a7b0422dc27459d8e51d0ed602688027cd5b 0 1531912348000 4 connected
0835372da4fa0056dc22c5d603dd6af3fbd7641c 192.168.248.144:8005@18005 slave 7a57fcf36ecc4c0fbaa30efd5545928abd3be81d 0 1531912344208 5 connected
619297db9a97533c4a5d8d1b57df535e2de5db9c 192.168.248.144:8002@18002 master - 0 1531912348294 2 connected 5461-10922
6a84a7b0422dc27459d8e51d0ed602688027cd5b 192.168.248.144:8003@18003 master - 0 1531912347276 3 connected 10923-16383
```

图 6-6

| 1 | `redis-server ../8007/redis.conf` |

启动成功后，进入 redisCluster 目录下，执行如下命令将该节点添加到集群中：

| 1 | `./redis-trib.rb add-node 192.168.248.144:8007 192.168.248.144:8001` |

中间的参数是要添加的 Redis 实例地址，最后的参数是集群中的实例。添加成功后，登录任意一个 Redis 实例，查看集群节点信息，就可以看到该实例已经被添加进集群了，如图 6-7 所示。

```
127.0.0.1:8001> cluster nodes
e0f2751b46c9ed3ca130e9fc825540386feaafb2 192.168.248.144:8007@18007 master - 0 1531914827000 0 connected
adc40fd1dd0007fe4b4f7aff47dc52ea8dbe1b7b 192.168.248.144:8006@18006 slave 619297db9a97533c4a5d8d1b57df535e2de5db9c 0 1531914829314 6 connected
7a57fcf36ecc4c0fbaa30efd5545928abd3be81d 192.168.248.144:8001@18001 myself,master - 0 1531914825000 1 connected 0-5460
75ec7962be8bc1d4b207e816458ba2c268d3329a 192.168.248.144:8004@18004 slave 6a84a7b0422dc27459d8e51d0ed602688027cd5b 0 1531914828293 4 connected
0835372da4fa0056dc22c5d603dd6af3fbd7641c 192.168.248.144:8005@18005 slave 7a57fcf36ecc4c0fbaa30efd5545928abd3be81d 0 1531914828000 5 connected
619297db9a97533c4a5d8d1b57df535e2de5db9c 192.168.248.144:8002@18002 master - 0 1531914830339 2 connected 5461-10922
6a84a7b0422dc27459d8e51d0ed602688027cd5b 192.168.248.144:8003@18003 master - 0 1531914829000 3 connected 10923-16383
```

图 6-7

可以看到，新实例已经被添加进集群中，但是由于 slot 已经被之前的实例分配完了，新添加的实例没有 slot，也就意味着新添加的实例没有存储数据的机会，此时需要从另外三个实例中拿出一部分 slot 分配给新实例，具体操作如下。

首先，在 redisCluster 目录下执行如下命令对 slot 重新分配：

```
1  ./redis-trib.rb reshard 192.168.248.144:8001
```

第二个参数表示连接集群中的任意一个实例。

在执行命令的过程中，有三个核心配置需要手动配置，如图 6-8 所示。

```
How many slots do you want to move (from 1 to 16384)? 1000
What is the receiving node ID? e0f2751b46c9ed3ca130e9fc825540386feaafb2
Please enter all the source node IDs.
  Type 'all' to use all the nodes as source nodes for the hash slots.
  Type 'done' once you entered all the source nodes IDs.
Source node #1:all
```

图 6-8

第一个配置是要拿出多少个 slot 分配给新实例，本案例配置了 1000 个。

第二个是把拿出来的 1000 个 slot 分配给谁，输入接收这 1000 个 slot 的 Redis 实例的 id，这个 id 在节点添加成功后就可以看到，也可以进入集群控制台后利用 cluster nodes 命令查看。

第三个配置是这 1000 个 slot 由哪个实例出，例如从端口为 8001 的实例中拿出 1000 个 slot 分配给端口为 8007 的实例，那么这里输入 8001 的 id 后按回车键，再输入 done 按回车键即可，如果想将 1000 个 slot 均摊到原有的所有实例中，那么这里输入 all 按回车键即可。

slot 分配成功后，再查看节点信息，就可以看到新实例也有 slot 了，如图 6-9 所示。

```
127.0.0.1:8001> cluster nodes
e0f2751b46c9ed3ca130e9fc825540386feaafb2 192.168.248.144:8007@18007 master - 0 1531915782000
0 7 connected 0-332 5461-5794 10923-11255
adc40fd1dd0007fe4b4f7ff47dc52ea8dbe1b7b 192.168.248.144:8006@18006 slave 619297db9a97533c4
a5d8d1b57df535e2de5db9c 0 1531915782000 6 connected
7a57fcf36ecc4c0fbaa30efd5545928abd3be81d 192.168.248.144:8001@18001 myself,master - 0 15319
15784000 1 connected 333-5460
75ec7962be8bc1d4b207e816458ba2c268d3329a 192.168.248.144:8004@18004 slave 6a84a7b0422dc2745
9d8e51d0ed602688027cd5b 0 1531915785617 4 connected
0835372da4fa0056dc22c5d603dd6af3fbd7641c 192.168.248.144:8005@18005 slave 7a57fcf36ecc4c0fb
aa30efd5545928abd3be81d 0 1531915783573 5 connected
619297db9a97533c4a5d8d1b57df535e2de5db9c 192.168.248.144:8002@18002 master - 0 153191578459
1 2 connected 5795-10922
6a84a7b0422dc27459d8e51d0ed602688027cd5b 192.168.248.144:8003@18003 master - 0 153191578255
5 3 connected 11256-16383
```

图 6-9

（6）添加从节点

上面添加的节点是主节点，从节点的添加相对要容易一些。添加从节点的步骤如下：

首先将 redisCluster 目录下的 8001 目录复制一份，命名为 8008，然后按照 6.1.2 小节中第 3 步的配置修改 8008 目录下的 redis.conf，修改完成后，启动该实例，然后输入如下命令添加从节点：

```
1  ./redis-trib.rb add-node --slave --master-id
   e0f2751b46c9ed3ca130e9fc825540386feaafb2 192.168.248.144:8008 192.168.248.144:8001
```

添加从节点需要指定该从节点的 masterid，--master-id 后面的参数即表示该从节点 master 的 id，192.168.248.144:8008 表示从节点的地址，192.168.248.144:8001 则表示集群中任意一个实例的地址。当从节点添加成功后，登录集群中任意一个 Redis 实例，通过 cluster nodes 命令就可以看到从节点的信息，如图 6-10 所示。

图 6-10

（7）删除节点

如果删除的是一个从节点，直接运行如下命令即可删除：

| 1 | ./redis-trib.rb del-node 192.168.248.144:8001 122b2098df746afc3a77beddaad85630bf75ab9a |

中间的实例地址表示集群中的任意一个实例，最后的参数表示要删除节点的 id。但若删除的节点占有 slot，则会删除失败，此时按照第 5 步提到的办法，先将要删除节点的 slot 全部都分配出去，然后运行如上命令就可以成功删除一个占有 slot 的节点了。

2．配置 Spring Boot

不同于单机版 Redis 整合 Spring Boot，Redis 集群整合 Spring Boot 需要开发者手动配置，配置步骤如下。

（1）创建 Spring Boot 项目

首先创建一个 Spring Boot Web 项目，添加如下依赖：

1	`<dependency>`
2	`<groupId>org.springframework.boot</groupId>`
3	`<artifactId>spring-boot-starter-web</artifactId>`
4	`</dependency>`
5	`<dependency>`
6	`<groupId>redis.clients</groupId>`
7	`<artifactId>jedis</artifactId>`
8	`</dependency>`
9	`<dependency>`
10	`<groupId>org.springframework.data</groupId>`
11	`<artifactId>spring-data-redis</artifactId>`
12	`</dependency>`

```
13    <dependency>
14        <groupId>org.apache.commons</groupId>
15        <artifactId>commons-pool2</artifactId>
16    </dependency>
```

（2）配置集群信息

由于集群节点有多个，可以保存在一个集合中，因此这里的配置文件使用 YAML 格式的，删除 resources 目录下的 application.properties 文件，创建 application.yml 配置文件（YAML 配置可以参考 2.7 节），文件内容如下：

```
1   spring:
2     redis:
3       cluster:
4         ports:
5           - 8001
6           - 8002
7           - 8003
8           - 8004
9           - 8005
10          - 8006
11          - 8007
12          - 8008
13        host: 192.168.248.144
14        poolConfig:
15          max-total: 8
16          max-idle: 8
17          max-wait-millis: -1
18          min-idle: 0
```

由于本案例 Redis 实例的 host 都是一样的，因此这里配置了一个 host，而 port 配置成了一个集合，这些 port 将被注入一个集合中。poolConfig 则是基本的连接池信息配置。

（3）配置 Redis

创建 RedisConfig，完成对 Redis 的配置，代码如下：

```
1   @Configuration
2   @ConfigurationProperties("spring.redis.cluster")
3   public class RedisConfig {
4       List<Integer> ports;
5       String host;
6       JedisPoolConfig poolConfig;
7       @Bean
8       RedisClusterConfiguration redisClusterConfiguration() {
9           RedisClusterConfiguration configuration = new RedisClusterConfiguration();
10          List<RedisNode> nodes = new ArrayList<>();
11          for (Integer port : ports) {
12              nodes.add(new RedisNode(host, port));
13          }
14          configuration.setPassword(RedisPassword.of("123@456"));
15          configuration.setClusterNodes(nodes);
16          return configuration;
```

```
17        }
18     @Bean
19     JedisConnectionFactory jedisConnectionFactory() {
20         JedisConnectionFactory factory = new
21     JedisConnectionFactory(redisClusterConfiguration(),poolConfig);
22         return factory;
23     }
24     @Bean
25     RedisTemplate redisTemplate() {
26         RedisTemplate redisTemplate = new RedisTemplate();
27         redisTemplate.setConnectionFactory(jedisConnectionFactory());
28         redisTemplate.setKeySerializer(new StringRedisSerializer());
29         redisTemplate.setValueSerializer(new JdkSerializationRedisSerializer());
30         return redisTemplate;
31     }
32     @Bean
33     StringRedisTemplate stringRedisTemplate() {
34         StringRedisTemplate stringRedisTemplate = new
35     StringRedisTemplate(jedisConnectionFactory());
36         stringRedisTemplate.setKeySerializer(new StringRedisSerializer());
37         stringRedisTemplate.setValueSerializer(new StringRedisSerializer());
38         return stringRedisTemplate;
39     }
40     //省略 getter/setter
41  }
```

代码解释：

- 通过@ConfigurationProperties 注解声明配置文件前缀，配置文件中定义的 ports 数组、host 以及连接池配置信息都将被注入 port、host、poolConfig 三个属性中。
- 配置 RedisClusterConfiguration 实例，设置 Redis 登录密码以及 Redis 节点信息。
- 根据 RedisClusterConfiguration 实例以及连接池配置信息创建 Jedis 连接工厂 JedisConnectionFactory。
- 根据 JedisConnectionFactory 创建 RedisTemplate 和 StringRedisTemplate，同时配置 key 和 value 的序列化方式。有了 RedisTemplate 和 StringRedisTemplate，剩下的用法就和单实例的用法一致了。

（4）创建 Controller

创建 Controller 和 Book 实例，代码如下：

```
1  public class Book implements Serializable {
2      private String name;
3      private String author;
4      //省略 getter/setter
5  }
6  @RestController
7  public class BookController {
8      @Autowired
9      RedisTemplate redisTemplate;
```

```
    @Autowired
    StringRedisTemplate stringRedisTemplate;
    @GetMapping("/test1")
    public void test1() {
        ValueOperations ops = redisTemplate.opsForValue();
        Book book = new Book();
        book.setName("水浒传");
        book.setAuthor("施耐庵");
        ops.set("b1", book);
        System.out.println(ops.get("b1"));
        ValueOperations<String, String> ops2 = stringRedisTemplate.opsForValue();
        ops2.set("k1", "v1");
        System.out.println(ops2.get("k1"));
    }
}
```

测试 Controller 与单实例 Redis 测试 Controller 基本一致。创建完成后，启动 Spring Boot 项目。

（5）测试

最后，在浏览器中输入 http://localhost:8080/test1，控制台打印日志如图 6-11 所示。

```
2018-07-18 22:39:05.878  INFO 13284 —
Book{name='水浒传', author='施耐庵'}
v1
```

图 6-11

然后登录任意一个 Redis 实例，查询数据，结果如图 6-12 所示。

```
192.168.248.144:8002> get k1
-> Redirected to slot [12706] located at 192.168.248.144:8003
"v1"
192.168.248.144:8003> get b1
-> Redirected to slot [2874] located at 192.168.248.144:8001
"\xac\xed\x00\x05sr\x00\rorg.sang.Book(\x19\x04\xf7|j\x8d\x12\x02\x00\x02L\x00\x06authort\x
00\x12Ljava/lang/String;L\x00\x04nameq\x00~\x00\x01xpt\x00\t\xe6\x96\xbd\xe8\x80\x90\xe5\xb
a\xb5t\x00\t\xe6\xb0\xb4\xe6\xb5\x92\xe4\xbc\xa0"
192.168.248.144:8001>
```

图 6-12

由图 6-12 的日志可以看到，查询时只需要登录任意一个 Redis 实例，RedisCluster 会负责将查询请求 Redirected 到相应的实例上去。

6.2　整合 MongoDB

6.2.1　MongoDB 简介

MongoDB 是一种面向文档的数据库管理系统，它是一个介于关系型数据库和非关系型数据库之间的产品，MongoDB 功能丰富，它支持一种类似 JSON 的 BSON 数据格式，既可以存储简单的数据格式，也可以存储复杂的数据类型。MongoDB 最大的特点是它支持的查询语言非常强大，并且还支持对数据建立索引。总体来说，MongoDB 是一款应用相当广泛的 NoSQL 数据库。

6.2.2 MongoDB 安装

本案例使用的 MongoDB 版本为写作本书时的最新版本 4.0.0，安装环境为 CentOS 7。安装步骤如下。

1. 下载 MongoDB

首先执行如下命令下载 MongoDB：

```
wget https://fastdl.mongodb.org/linux/mongodb-linux-x86_64-4.0.0.tgz
```

下载完成后，将下载的 MongoDB 解压，并将解压后的文件夹重命名为 mongodb，执行命令如下：

```
tar -zxvf mongodb-linux-x86_64-4.0.0.tgz
mv mongodb-linux-x86_64-4.0.0 mongodb
```

2. 配置 MongoDB

进入 mongodb 目录下，创建两个文件夹 db 和 logs，分别用来保存数据和日志，代码如下：

```
cd mongodb
mkdir db
mkdir logs
```

然后进入 bin 目录下，创建一个新的 MongoDB 配置文件 mongo.conf，文件内容如下：

```
dbpath=/opt/mongodb/db
logpath=/opt/mongodb/logs/mongodb.log
port=27017
fork=true
```

配置解释：

- 第 1 行配置表示数据存储目录。
- 第 2 行配置表示日志文件位置。
- 第 3 行配置表示启动端口。
- 第 4 行配置表示以守护程序的方式启动 MongoDB，即允许 MongoDB 在后台运行。

3. MongoDB 的启动和关闭

配置完成后，还是在 bin 目录下，运行如下命令启动 MongoDB：

```
./mongod -f mongo.conf --bind_ip_all
```

-f 表示指定配置文件的位置，--bind_ip_all 则表示允许所有的远程地址连接该 MongoDB 实例。MongoDB 启动成功后，在 bin 目录下再执行 mongo 命令，进入 MongoDB 控制台，然后输入

db.version(),如果能看到 MongoDB 的版本号,就表示安装成功:

```
1  ./mongo
2  db.version()
```

安装成功的界面如图 6-13 所示。

图 6-13

默认情况下,启动后连接的是 MongoDB 中的 test 库,而关闭 MongoDB 的命令需要在 admin 库中执行,因此关闭 MongoDB 需要首先切换到 admin 库,然后执行 db.shutdownServer();命令,完整操作步骤如下:

```
1  use admin;
2  db.shutdownServer();
3  exit
```

服务关闭后,执行 exit 命令退出控制台,此时如果再执行./mongo 命令就会执行失败,如图 6-14 所示。

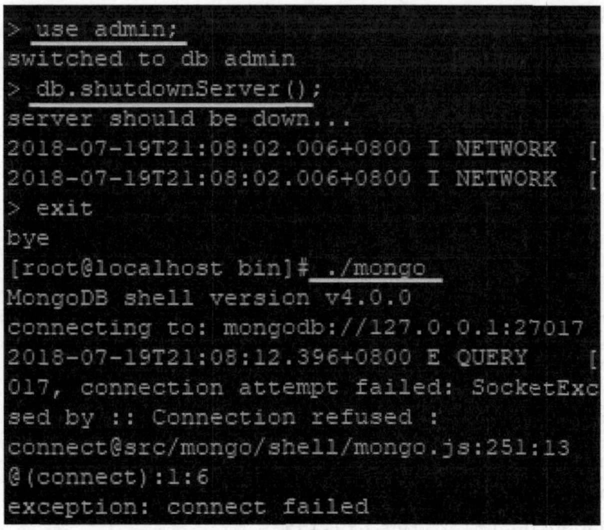

图 6-14

4. 安全管理

默认情况下,启动的 MongoDB 没有登录密码,在生产环境中这是非常不安全的,但是不同于 MySQL、Oracle 等关系型数据库,MongoDB 中每一个库都有独立的密码,在哪一个库中创建用户就要在哪一个库中验证密码。要配置密码,首先要创建一个用户,例如在 admin 库中创建一个用户,代码如下:

```
1  use admin;
2  db.createUser({user:"sang",pwd:"123",roles:[{role:"readWrite",db:"test"}]})
```

新创建的用户名为 sang，密码是 123，roles 表示该用户具有的角色，这里的配置表示该用户对 test 库具有读和写两项权限。

用户创建成功后，关闭当前实例，然后重新启动，启动命令如下：

```
1  ./mongod -f mongo.conf --auth --bind_ip_all
```

启动成功后，再次进入控制台，然后切换到 admin 库中验证登录（默认连接上的库是 test 库），验证成功后就可以对 test 库执行读写操作了，代码如下：

```
1  ./mongo
2  db.auth("sang","123")
```

如果 db.auth("sang","123") 命令执行结果为 1，就表示认证成功，可以执行对 test 库的读写操作。

6.2.3 整合 Spring Boot

借助于 Spring Data MongoDB，Spring Boot 为 MongoDB 也提供了开箱即用的自动化配置方案，具体配置步骤如下。

1. 创建 Spring Boot 工程

创建 Spring Boot Web 工程，添加 MongoDB 依赖，代码如下：

```
1  <dependency>
2    <groupId>org.springframework.boot</groupId>
3    <artifactId>spring-boot-starter-data-mongodb</artifactId>
4  </dependency>
5  <dependency>
6    <groupId>org.springframework.boot</groupId>
7    <artifactId>spring-boot-starter-web</artifactId>
8  </dependency>
```

2. 配置 MongoDB

在 application.properties 中配置 MongoDB 的连接信息，代码如下：

```
1  spring.data.mongodb.authentication-database=admin
2  spring.data.mongodb.database=test
3  spring.data.mongodb.host=192.168.248.144
4  spring.data.mongodb.port=27017
5  spring.data.mongodb.username=sang
6  spring.data.mongodb.password=123
```

配置解释：

- 第 1 行配置表示验证登录信息的库。
- 第 2 行配置表示要连接的库，认证信息不一定要在连接的库中创建，因此这两个分开配置。
- 第 3~6 行配置表示 MongoDB 的连接地址和认证信息等。

3. 创建实体类

创建实体类 Book，代码如下：

```
public class Book{
    private Integer id;
    private String name;
    private String author;
    //省略getter/setter
}
```

4. 创建 BookDao

BookDao 的定义类似于 Spring Data JPA 中的 Repository 定义，代码如下：

```
public interface BookDao extends MongoRepository<Book,Integer> {
    List<Book> findByAuthorContains(String author);
    Book findByNameEquals(String name);
}
```

MongoRepository 中已经预定义了针对实体类的查询、添加、删除等操作。BookDao 中可以按照 5.3 节提到的方法命名规则定义查询方法。

5. 创建 Controller

简单起见，直接将 BookDao 注入 Controller 进行测试：

```
@RestController
public class BookController {
    @Autowired
    BookDao bookDao;
    @GetMapping("/test1")
    public void test1() {
        List<Book> books = new ArrayList<>();
        Book b1 = new Book();
        b1.setId(1);
        b1.setName("朝花夕拾");
        b1.setAuthor("鲁迅");
        books.add(b1);
        Book b2 = new Book();
        b2.setId(2);
        b2.setName("呐喊");
        b2.setAuthor("鲁迅");
        books.add(b2);
        bookDao.insert(books);
        List<Book> books1 = bookDao.findByAuthorContains("鲁迅");
        System.out.println(books1);
        Book book = bookDao.findByNameEquals("朝花夕拾");
        System.out.println(book);
    }
}
```

代码解释：

- 第 18 行调用 MongoRepository 中的 insert 方法插入集合中的数据。
- 第 19 行表示查询作者名字中包含"鲁迅"的所有书。
- 第 21 行表示查询书名为"朝花夕拾"的图书信息。

6. 测试 BookDao

创建好 Controller 后，在浏览器中输入 http://localhost:8080/test1，控制台打印日志如图 6-15 所示。

```
[Book{id=1, name='朝花夕拾', author='鲁迅'}, Book{id=2, name='呐喊', author='鲁迅'}]
Book{id=1, name='朝花夕拾', author='鲁迅'}
```

图 6-15

此时登录 MongoDB 服务器，认证身份后，在 test 库中即可查询到刚刚插入的数据，如图 6-16 所示。

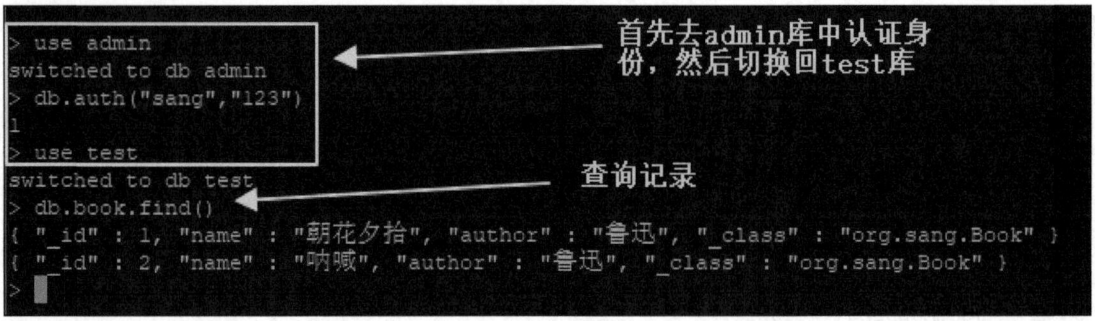

```
> use admin
switched to db admin
> db.auth("sang","123")
1
> use test
switched to db test
> db.book.find()
{ "_id" : 1, "name" : "朝花夕拾", "author" : "鲁迅", "_class" : "org.sang.Book" }
{ "_id" : 2, "name" : "呐喊", "author" : "鲁迅", "_class" : "org.sang.Book" }
>
```

首先去admin库中认证身份，然后切换回test库

查询记录

图 6-16

7. 使用 MongoTemplate

除了继承 MongoRepository 外，Spring Data MongoDB 还提供了 MongoTemplate 用来方便地操作 MongoDB。在 Spring Boot 中，若添加了 MongoDB 相关的依赖，而开发者并没有提供 MongoTemplate，则默认会有一个 MongoTemplate 注册到 Spring 容器中，相关配置源码在 MongoDataAutoConfiguration 类中。因此，用户可以直接使用 MongoTemplate，在 Controller 中直接注入 MongoTemplate 就可以使用了，添加如下代码到第 5 步的 Controller 中：

```
1   @Autowired
2   MongoTemplate mongoTemplate;
3   @GetMapping("/test2")
4   public void test2() {
5       List<Book> books = new ArrayList<>();
6       Book b1 = new Book();
7       b1.setId(3);
8       b1.setName("围城");
9       b1.setAuthor("钱钟书");
10      books.add(b1);
11      Book b2 = new Book();
```

```
12          b2.setId(4);
13          b2.setName("宋诗选注");
14          b2.setAuthor("钱钟书");
15          books.add(b2);
16          mongoTemplate.insertAll(books);
17          List<Book> list = mongoTemplate.findAll(Book.class);
18          System.out.println(list);
19          Book book = mongoTemplate.findById(3, Book.class);
20          System.out.println(book);
21      }
```

代码解释：

- 第 1、2 行表示注入 Spring Boot 提供的 MongoTemplate。
- 第 16 行表示向 MongoDB 中插入一个集合。
- 第 17 行表示查询 book 集合中的所有数据。
- 第 19 行表示根据 id 查询一个文档。

最后，在浏览器中输入 http://localhost:8080/test2，控制台打印日志如图 6-17 所示。

```
[Book{id=1, name='朝花夕拾', author='鲁迅'}, Book{id=2, name='呐喊', author='鲁迅'},
 Book{id=3, name='围城', author='钱钟书'}, Book{id=4, name='宋诗选注', author='钱钟书'}]
Book{id=3, name='围城', author='钱钟书'}
```

图 6-17

6.3 Session 共享

正常情况下，HttpSession 是通过 Servlet 容器创建并进行管理的，创建成功之后都是保存在内存中。如果开发者需要对项目进行横向扩展搭建集群，那么可以利用一些硬件或者软件工具来做负载均衡，此时，来自同一用户的 HTTP 请求就有可能被分发到不同的实例上去，如何保证各个实例之间 Session 的同步就成为一个必须解决的问题。Spring Boot 提供了自动化的 Session 共享配置，它结合 Redis 可以非常方便地解决这个问题。使用 Redis 解决 Session 共享问题的原理非常简单，就是把原本存储在不同服务器上的 Session 拿出来放在一个独立的服务器上，如图 6-18 所示。

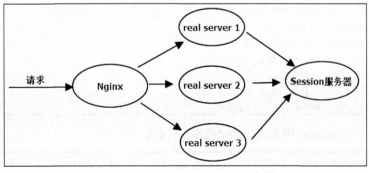

图 6-18

当一个请求到达 Nginx 服务器后，首先进行请求分发，假设请求被 real server1 处理了，real server 1 在处理请求时，无论是存储 Session 还是读取 Session，都去操作 Session 服务器而不是操作自身内存中的 Session，其他 real server 在处理请求时也是如此，这样就可以实现 Session 共享了。

6.3.1　Session 共享配置

Spring Boot 中的 Session 共享配置非常容易，创建 Spring Boot Web 项目，添加 Redis 和 Session 依赖，代码如下：

```xml
1   <dependency>
2     <groupId>org.springframework.boot</groupId>
3     <artifactId>spring-boot-starter-data-redis</artifactId>
4     <exclusions>
5       <exclusion>
6         <groupId>io.lettuce</groupId>
7         <artifactId>lettuce-core</artifactId>
8       </exclusion>
9     </exclusions>
10  </dependency>
11  <dependency>
12    <groupId>redis.clients</groupId>
13    <artifactId>jedis</artifactId>
14  </dependency>
15  <dependency>
16    <groupId>org.springframework.boot</groupId>
17    <artifactId>spring-boot-starter-web</artifactId>
18  </dependency>
19  <dependency>
20    <groupId>org.springframework.session</groupId>
21    <artifactId>spring-session-data-redis</artifactId>
22  </dependency>
```

除了 Redis 依赖之外，这里还要提供 spring-session-data-redis 依赖，Spring Session 可以做到透明化地替换掉应用的 Session 容器。项目创建成功后，在 application.properties 中进行 Redis 基本连接信息配置，代码如下：

```
1   spring.redis.database=0
2   spring.redis.host=192.168.66.130
3   spring.redis.port=6379
4   spring.redis.password=123@456
5   spring.redis.jedis.pool.max-active=8
6   spring.redis.jedis.pool.max-idle=8
7   spring.redis.jedis.pool.max-wait=-1ms
8   spring.redis.jedis.pool.min-idle=0
```

然后创建一个 Controller 用来执行测试操作，代码如下：

```java
1   @RestController
2   public class HelloController {
```

```
3      @Value("${server.port}")
4      String port;
5      @PostMapping("/save")
6      public String saveName(String name, HttpSession session) {
7          session.setAttribute("name", name);
8          return port;
9      }
10     @GetMapping("/get")
11     public String getName(HttpSession session) {
12         return port + ":" + session.getAttribute("name").toString();
13     }
14 }
```

这里提供了两个方法,一个 save 接口用来向 Session 中存储数据,还有一个 get 接口用来从 Session 中获取数据,这里注入了项目启动的端口号 server.port,主要是为了区分到底是哪个服务器提供的服务。另外,虽然还是操作的 HttpSession,但是实际上 HttpSession 容器已经被透明替换,真正的 Session 此时存储在 Redis 服务器上。

项目创建完成后,将项目打成 jar 包上传到 CentOS 上。然后执行如下两条命令启动项目:

```
1  nohup java -jar session-0.0.1-SNAPSHOT.jar --server.port=8080 &
2  nohup java -jar session-0.0.1-SNAPSHOT.jar --server.port=8081 &
```

nohup 表示不挂断程序运行,即当终端窗口关闭后,程序依然在后台运行,最后的&表示让程序在后台运行。--server.port 表示设置启动端口,一个为 8080,另一个为 8081。启动成功后,接下来就可以配置负载均衡器了。(关于 Linux 上如何运行 Spring Boot 工程可以参考本书第 15 章。)

6.3.2 Nginx 负载均衡

本案例使用 Nginx 做负载均衡。首先在 CentOS 上安装 Nginx,安装过程如下。

下载源码并解压:

```
1  wget https://nginx.org/download/nginx-1.14.0.tar.gz
2  tar -zxvf nginx-1.14.0.tar.gz
```

然后进入解压目录中执行编译安装,代码如下:

```
1  cd nginx-1.14.0
2  ./configure
3  make
4  make install
```

安装成功后,找到 Nginx 安装目录,执行 sbin 目录下的 nginx 文件启动 nginx,命令如下:

```
1  /usr/local/nginx/sbin/nginx
```

Nginx 启动成功后,默认端口是 80,可以在物理机直接访问,如图 6-19 所示。

图 6-19

接下来进入 Nginx 安装目录修改配置文件，代码如下：

```
1  vi /usr/local/nginx/conf/nginx.conf
```

对 nginx.conf 文件进行编辑，编辑内容如下：

```
1  …
2  …
3  upstream sang.com{
4    server 192.168.66.130:8080 weight=1;
5    server 192.168.66.130:8081 weight=1;
6  }
7  server {
8      listen       80;
9      server_name  localhost;
10     location / {
11      proxy_pass http://sang.com;
12       proxy_redirect default;
13     }
14  …
15  …
16  }
```

> **注 意**
>
> 这里只列出了修改的配置，在修改的配置中首先配置上游服务器，即两个 real server，两个 real server 的权重都是 1，意味着请求将平均分配到两个 real server 上，然后在 server 中配置拦截规则，将拦截到的请求转发到定义好的 real server 上。

配置完成后，重启 Nginx，重启命令如下：

```
1  /usr/local/nginx/sbin/nginx -s reload
```

6.3.3 请求分发

当 real server 和 Nginx 都启动后，调用 "/save" 接口存储数据，如图 6-20 所示。

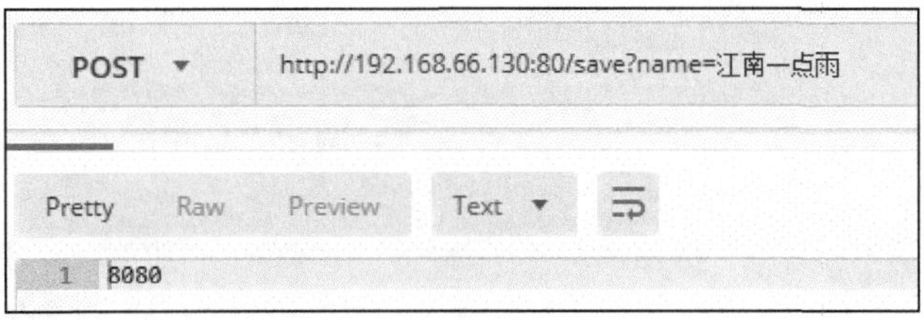

第 6 章 Spring Boot 整合 NoSQL

图 6-20

> **注　意**
>
> 调用的端口是 80，即调用的是 Nginx 服务器，请求会被 Nginx 转发到 real server 上进行处理，返回值为 8080，说明真正处理请求的 real server 是 8080 那台服务器，接下来调用 get 接口获取数据，如图 6-21 所示。

图 6-21

> **注　意**
>
> 调用端口依然是 80，但是返回值是 8081，说明是 8081 那台 real server 提供的服务，如果这里不是 8081，再访问一次即可。

经过如上步骤，就完成了利用 Redis 实现 Session 共享的功能，基本上不需要额外配置，开箱即用。

6.4 小　结

本章主要向读者介绍了 Spring Boot 整合 NoSQL 数据库以及结合 Redis 实现 Session 共享。对于 NoSQL 数据库，介绍了比较常见的两种：MongoDB 和 Redis。MongoDB 在一些场景中甚至可以完全替代关系型数据库，Redis 更多的使用场景则是作为缓存服务器（本书第 9 章将详细介绍 Redis 缓存），开发者可根据具体情况选择合适的 NoSQL。

第 7 章

构建 RESTful 服务

本章概要

- REST 简介
- JPA 实现 REST
- MongoDB 实现 REST

7.1 REST 简介

REST（Representational State Transfer）是一种 Web 软件架构风格，它是一种风格，而不是标准，匹配或兼容这种架构风格的网络服务称为 REST 服务。REST 服务简洁并且有层次，REST 通常基于 HTTP、URI 和 XML 以及 HTML 这些现有的广泛流行的协议和标准。在 REST 中，资源是由 URI 来指定的，对资源的增删改查操作可以通过 HTTP 协议提供的 GET、POST、PUT、DELETE 等方法实现。使用 REST 可以更高效地利用缓存来提高响应速度，同时 REST 中的通信会话状态由客户端来维护，这可以让不同的服务器处理一系列请求中的不同请求，进而提高服务器的扩展性。在前后端分离项目中，一个设计良好的 Web 软件架构必然要满足 REST 风格。

在 Spring MVC 框架中，开发者可以通过@RestController 注解开发一个 RESTful 服务，不过，Spring Boot 对此提供了自动化配置方案，开发者只需要添加相关依赖就能快速构建一个 RESTful 服务。

7.2 JPA 实现 REST

在 Spring Boot 中，使用 Spring Data JPA 和 Spring Data Rest 可以快速开发出一个 RESTful 应用。接下来向读者介绍 Spring Boot 中非常方便的 RESTful 应用开发。

7.2.1 基本实现

1．创建项目

创建 Spring Boot 项目，添加如下依赖：

```
1   <dependency>
2     <groupId>org.springframework.boot</groupId>
3     <artifactId>spring-boot-starter-data-jpa</artifactId>
4   </dependency>
5   <dependency>
6     <groupId>org.springframework.boot</groupId>
7     <artifactId>spring-boot-starter-data-rest</artifactId>
8   </dependency>
9   <dependency>
10    <groupId>com.alibaba</groupId>
11    <artifactId>druid</artifactId>
12    <version>1.1.9</version>
13  </dependency>
14  <dependency>
15    <groupId>mysql</groupId>
16    <artifactId>mysql-connector-java</artifactId>
17    <scope>runtime</scope>
18  </dependency>
```

这里的依赖除了数据库相关的依赖外，还有 Spring Data JPA 的依赖以及 Spring Data Rest 的依赖。项目创建完成后，在 application.properties 中配置基本的数据库连接信息：

```
1   spring.datasource.type=com.alibaba.druid.pool.DruidDataSource
2   spring.datasource.username=root
3   spring.datasource.password=123
4   spring.datasource.url=jdbc:mysql:///jparestful
5   spring.jpa.hibernate.ddl-auto=update
6   spring.jpa.database=mysql
7   spring.jpa.properties.hibernate.dialect=org.hibernate.dialect.MySQL57Dialect
8   spring.jpa.show-sql=true
```

这些数据库配置和 5.3 节中配置 JPA 的信息基本一致，配置的含义就不再赘述了。

2．创建实体类

```
1   @Entity(name = "t_book")
```

```
2    public class Book {
3        @Id
4        @GeneratedValue(strategy = GenerationType.IDENTITY)
5        private Integer id;
6        private String name;
7        private String author;
8        //省略 getter/setter
9    }
```

3. 创建 BookRepository

```
1    public interface BookRepository extends JpaRepository<Book, Integer> {
2    }
```

创建 BookRepository 类继承 JpaRepository，JpaRepository 中默认提供了一些基本的操作方法，代码如下：

```
1    @NoRepositoryBean
2    public interface JpaRepository<T, ID> extends PagingAndSortingRepository<T, ID>,
3                    QueryByExampleExecutor<T> {
4        List<T> findAll();
5        List<T> findAll(Sort sort);
6        List<T> findAllById(Iterable<ID> ids);
7        <S extends T> List<S> saveAll(Iterable<S> entities);
8        void flush();
9        <S extends T> S saveAndFlush(S entity);
10       void deleteInBatch(Iterable<T> entities);
11       void deleteAllInBatch();
12       T getOne(ID id);
13       @Override
14       <S extends T> List<S> findAll(Example<S> example);
15       @Override
16       <S extends T> List<S> findAll(Example<S> example, Sort sort);
17   }
```

由这段源码可以看到，基本的增删改查、分页查询方法 JpaRepository 都提供了。

4．测试

经过如上几步，一个 RESTful 服务就构建成功了，可能有读者会问"什么都没写呀！"，是的，这就是 Spring Boot 的魅力所在。

RESTful 的测试首先需要有一个测试工具，可以直接使用浏览器中的插件，例如 Firefox 中的 RESTClient，或者直接使用 Postman 等工具，笔者这里是使用 Postman 测试的。

5．添加测试

RESTful 服务构建成功后，默认的请求路径是实体类名小写再加上后缀。

此时向数据库添加一条数据非常容易，发起一个 post 请求，请求地址为 http://localhost:8080/books，如图 7-1 所示。

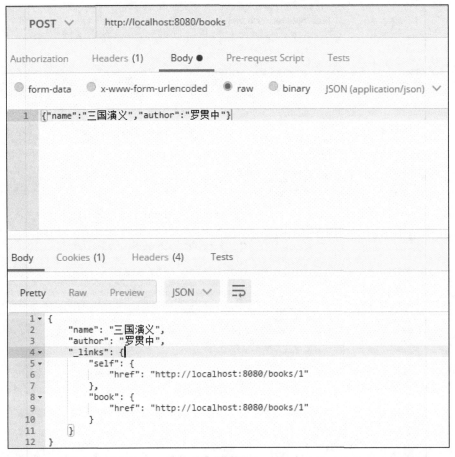

图 7-1

当添加成功后，服务端会返回刚刚添加成功的数据的基本信息以及浏览地址。

6．查询测试

查询是 GET 请求，分页查询请求路径为/books，请求 URL 如下：

```
1  http://localhost:8080/books
```

分页查询请求默认每页记录数是 20 条，页数为 0（页码从 0 开始计），查询结果如图 7-2 所示。

如果按照 id 查询，只需要在/books 后面追加上 id 即可（如图 7-3 所示），例如查询 id 为 1 的 book，请求 URL 如下：

```
1  http://localhost:8080/books/1
```

```
GET        http://localhost:8080/books

Body    Cookies (1)   Headers (3)   Tests

Pretty   Raw   Preview   JSON

 1 - {
 2      "_embedded": {
 3          "books": [
 4 -            {
 5                  "name": "三国演义",
 6                  "author": "罗贯中",
 7 -                "_links": {
 8 -                    "self": {
 9                          "href": "http://localhost:8080/books/1"
10                      },
11 -                    "book": {
12                          "href": "http://localhost:8080/books/1"
13                      }
14                  }
15              },
16 -            {
17                  "name": "红楼梦",
18                  "author": "曹雪芹",
19 -                "_links": {
20 -                    "self": {
21                          "href": "http://localhost:8080/books/2"
22                      },
23 -                    "book": {
24                          "href": "http://localhost:8080/books/2"
25                      }
26                  }
27              }
28          ]
29      },
30 -    "_links": {
31 -        "self": {
32              "href": "http://localhost:8080/books{?page,size,sort}",
33              "templated": true
34          },
35 -        "profile": {
36              "href": "http://localhost:8080/profile/books"
37          }
38      },
39 -    "page": {
40          "size": 20,
41          "totalElements": 2,
42          "totalPages": 1,
43          "number": 0
44      }
45 }
```

图 7-2

在图 7-2 查询所有数据返回的结果中，除了所有图书的基本信息外，还有如何发起一个分页请求以及当前页面的分页信息。如果开发者想要修改请求页码和每页记录数，只需要在请求地址中携带上相关参数即可，如下请求表示查询第 2 页数据并且每页记录数为 3：

```
1  http://localhost:8080/books?page=1&size=3
```

除了分页外，默认还支持排序，例如想查询第 2 页数据，每页记录数为 3，并且按照 id 倒序排列，请求地址如下：

```
1  http://localhost:8080/books?page=1&size=3&sort=id,desc
```

图 7-3

7. 修改测试

发送 PUT 请求可实现对数据的修改，对数据的修改是通过 id 进行的，因此请求路径中要有 id，例如如下请求路径表示修改 id 为 2 的记录，具体的修改内容在请求体中，如图 7-4 所示。

```
1  http://localhost:8080/books/2
```

图 7-4

PUT 请求的返回结果就是被修改之后的记录。

8. 删除测试

发送 DELETE 请求可以实现对数据的删除操作，例如删除 id 为 1 的记录，请求 URL 如下：

```
http://localhost:8080/books/1
```

DELETE 请求没有返回值，上面这个请求发送成功后，id 为 1 的记录就被删除了。

7.2.2 自定义请求路径

默认情况下，请求路径都是实体类名小写加 s，如果开发者想对请求路径进行重定义，通过 @RepositoryRestResource 注解即可实现，下面的案例只需在 BookRepository 上添加 @RepositoryRestResource 注解即可：

```
@RepositoryRestResource(path = "bs",collectionResourceRel = "bs",itemResourceRel = "b")
public interface BookRepository extends JpaRepository<Book, Integer> {
}
```

@RepositoryRestResource 注解的 path 属性表示将所有请求路径中的 books 都修改为 bs，如 http://localhost:8080/bs；collectionResourceRel 属性表示将返回的 JSON 集合中 book 集合的 key 修改为 bs；itemResourceRel 表示将返回的 JSON 集合中的单个 book 的 key 修改为 b，如图 7-5 所示。

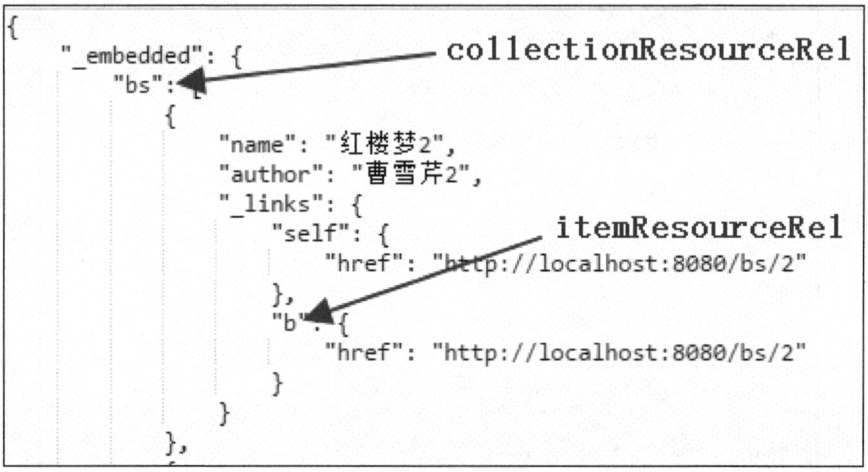

```
{
    "_embedded": {
        "bs":                      ← collectionResourceRel
        {
            "name": "红楼梦2",
            "author": "曹雪芹2",
            "_links": {
                "self": {
                    "href": "http://localhost:8080/bs/2"
                },              ← itemResourceRel
                "b": {
                    "href": "http://localhost:8080/bs/2"
                }
            }
        },
```

图 7-5

7.2.3 自定义查询方法

默认的查询方法支持分页查询、排序查询以及按照 id 查询，如果开发者想要按照某个属性查询，只需在 BookRepository 中定义相关方法并暴露出去即可，代码如下：

```
@RepositoryRestResource(path = "bs",collectionResourceRel = "bs",itemResourceRel = "b")
```

```
3    public interface BookRepository extends JpaRepository<Book, Integer> {
4        @RestResource(path = "author",rel = "author")
5        List<Book> findByAuthorContains(@Param("author") String author);
6        @RestResource(path = "name",rel = "name")
7        Book findByNameEquals(@Param("name") String name);
     }
```

代码解释：

- 自定义查询只需要在 BookRepository 中定义相关查询方法即可，方法定义好之后可以不添加 @RestResource 注解，默认路径就是方法名。以第 4 行定义的方法为例，若不添加 @RestResource 注解，则默认该方法的调用路径为 http://localhost:8080/bs/search/findByAuthorContains?author=鲁迅。如果想对查询路径进行自定义，只需要添加@RestResource 注解即可，path 属性即表示最新的路径。还是以第 4 行的方法为例，添加 @RestResource(path = "author",rel = "author") 注解后的查询路径为 "http://localhost:8080/bs/search/author?author=鲁迅"。
- 用户可以直接访问 http://localhost:8080/bs/search 路径查看该实体类暴露出来了哪些查询方法，默认情况下，在查询方法展示时使用的路径是方法名，通过@RestResource 注解中的 rel 属性可以对这里的路径进行重定义，如图 7-6 所示。

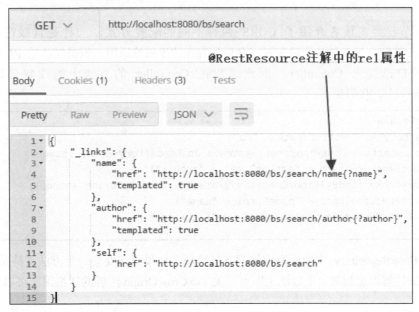

图 7-6

7.2.4　隐藏方法

默认情况下，凡是继承了 Repository 接口（或者 Repository 的子类）的类都会被暴露出来，即开发者可执行基本的增删改查方法。以上文的 BookRepository 为例，如果开发者提供了 BookRepository 继承自 Repository，就能执行对 Book 的基本操作，如果开发者继承了 Repository

但是又不想暴露相关操作，做如下配置即可：

```
1  @RepositoryRestResource(exported = false,)
2  public interface BookRepository extends JpaRepository<Book, Integer> {
3  }
```

将@RepositoryRestResource 注解中的 exported 属性置为 false 之后，则 7.2.4 小节中展示的增删改查接口都会失效，BookRepository 类中定义的相关方法也会失效。若只是单纯地不想暴露某个方法，则在方法上进行配置即可，例如开发者想屏蔽 DELETE 接口，做如下配置即可：

```
1  @RepositoryRestResource(path = "bs",collectionResourceRel = "bs",itemResourceRel =
2  "b")
3  public interface BookRepository extends JpaRepository<Book, Integer> {
4      @Override
5      @RestResource(exported = false)
6      void deleteById(Integer integer);
7  }
```

@RestResource 注解的 exported 属性默认为 true，将之置为 false 即可。

7.2.5 配置 CORS

在 4.6 节已经向读者介绍了 CORS 两种不同的配置方式，一种是直接在方法上添加 @CrossOrigin 注解，另一种是全局配置。全局配置在这里依然适用，但是默认的 RESTful 工程不需要开发者自己提供 Controller，因此添加在 Controller 的方法上的注解可以直接写在 BookRepository 上，代码如下：

```
1  @CrossOrigin
2  @RepositoryRestResource(path = "bs")
3  public interface BookRepository extends JpaRepository<Book, Integer> {
4      @RestResource(path = "author",rel = "author")
5      List<Book> findByAuthorContains(@Param("author") String author);
6      @RestResource(path = "name",rel = "name")
7      Book findByNameEquals(@Param("name") String name);
8  }
```

此时，BookRepository 中的所有方法都支持跨域。如果只需要某一个方法支持跨域，那么将 @CrossOrigin 注解添加到某一个方法上即可。关于@CrossOrigin 注解的详细用法可以参考 4.6 节。

7.2.6 其他配置

开发者也可以在 application.properties 中配置一些常用属性，代码如下：

```
1  #每页默认记录数，默认值为 20
2  spring.data.rest.default-page-size=2
3  #分页查询页码参数名，默认值为 page
4  spring.data.rest.page-param-name=page
5  #分页查询记录数参数名，默认值为 size
```

```
6   spring.data.rest.limit-param-name=size
7   #分页查询排序参数名,默认值为 sort
8   spring.data.rest.sort-param-name=sort
9   #base-path 表示给所有请求路径都加上前缀
10  spring.data.rest.base-path=/api
11  #添加成功时是否返回添加内容
12  spring.data.rest.return-body-on-create=true
13  #更新成功时是否返回更新内容
14  spring.data.rest.return-body-on-update=true
```

当然,这些 XML 配置也可以在 Java 代码中配置,且代码中配置的优先级高于 application.properties 配置的优先级,代码如下:

```
1   @Configuration
2   public class RestConfig extends RepositoryRestConfigurerAdapter {
3       @Override
4       public void configureRepositoryRestConfiguration(RepositoryRestConfiguration
5   config) {
6           config.setDefaultPageSize(2)
7                   .setPageParamName("page")
8                   .setLimitParamName("size")
9                   .setSortParamName("sort")
10                  .setBasePath("/api")
11                  .setReturnBodyOnCreate(true)
12                  .setReturnBodyOnUpdate(true);
13      }
    }
```

这里每项代码配置的含义都和 application.properties 中的配置一一对应,因此不再赘述。

7.3 MongoDB 实现 REST

在 6.2 节中向读者介绍了 MongoDB 整合 Spring Boot,而使用 Spring Boot 快速构建 RESTful 服务除了结合 Spring Data JPA 之外,也可以结合 Spring Data MongoDB 实现。使用 Spring Data MongoDB 构建 RESTful 服务也是三个步骤,分别如下。

1. 创建项目

首先创建 Spring Boot Web 项目,添加如下依赖:

```
1   <dependency>
2       <groupId>org.springframework.boot</groupId>
3       <artifactId>spring-boot-starter-data-mongodb</artifactId>
4   </dependency>
5   <dependency>
6       <groupId>org.springframework.boot</groupId>
7       <artifactId>spring-boot-starter-data-rest</artifactId>
8   </dependency>
```

这里 Spring Data Rest 的依赖和 7.2 节中的一致，只是将 Spring Data JPA 的依赖变为 Spring Data MongoDB 的依赖。项目创建成功后，在 application.properties 中配置 MongoDB 的基本连接信息，代码如下：

```
1  spring.data.mongodb.authentication-database=test
2  spring.data.mongodb.database=test
3  spring.data.mongodb.username=sang
4  spring.data.mongodb.password=123
5  spring.data.mongodb.host=192.168.248.144
6  spring.data.mongodb.port=27017
```

这段配置的含义可以参考 6.2.3 节，这里不再赘述。

2. 创建实体类

接下来创建一个普通的 Book 实体类，代码如下：

```
1  public class Book {
2      private Integer id;
3      private String name;
4      private String author;
5      //省略 getter/setter
6  }
```

3. 创建 BookRepository

创建 BookRepository 实现对 Book 的基本操作：

```
1  public interface BookRepository extends MongoRepository<Book,Integer> {
2  }
```

如此之后，一个 RESTful 服务就搭建成功了。在启动 Spring Boot 项目之前，记得要先启动 MongoDB。Spring Boot 项目启动成功后，接下来的测试环节与 7.2.1 小节的第 5~8 步一致。另外，7.2.2~7.2.6 小节介绍的 Spring Data Rest 配置在这里一样适用，因此不再赘述。

7.4 小　　结

本章向读者介绍了 Spring Boot 构建 RESTful 服务，结合 Spring Data Rest、Spring Data JPA 以及 Spring Data MongoDB，Spring Boot 可以快速构建出一个基本的 RESTful 服务，而开发者可以结合具体情况选择关系型数据库或者非关系型数据库作为数据支撑。在一些常规功能的项目中，Spring Boot 的这些特性可以帮助开发者省去许多繁杂臃肿的配置。

第 8 章

开发者工具与单元测试

本章概要

- devtools 简介
- devtools 实战
- 单元测试

8.1 devtools 简介

Spring Boot 中提供了一组开发工具 spring-boot-devtools，可以提高开发者的工作效率，开发者可以将该模块包含在任何项目中，spring-boot-devtools 最方便的地方莫过于热部署了。

8.2 devtools 实战

8.2.1 基本用法

要想在项目中加入 devtools 模块，只需添加相关依赖即可，代码如下：

```
1  <dependency>
2    <groupId>org.springframework.boot</groupId>
```

3	`<artifactId>spring-boot-devtools</artifactId>`
4	`<optional>true</optional>`
5	`</dependency>`

> **注　意**
>
> 这里多了一个 optional 选项，是为了防止将 devtools 依赖传递到其他模块中。当开发者将应用打包运行后，devtools 会被自动禁用。

当开发者将 spring-boot-devtools 引入项目后，只要 classpath 路径下的文件发生了变化，项目就会自动重启，这极大地提高了项目的开发速度。如果开发者使用了 Eclipse，那么在修改完代码并保存之后，项目将自动编译并触发重启，而开发如果使用了 IntelliJ IDEA，默认情况下，需要开发者手动编译才会触发重启。手动编译时，单击 Build→Build Project 菜单或者按 Ctrl+F9 快捷键进行编译，编译成功后就会触发项目重启。当然，使用 IntelliJ IDEA 的开发者也可以配置项目自动编译，配置步骤如下：

步骤01 单击 File→Settings 菜单，打开 Settings 页面，在左边的菜单栏依次找到 Build,Execution,Deployment→Compile，勾选 Build project automatically，如图 8-1 所示。

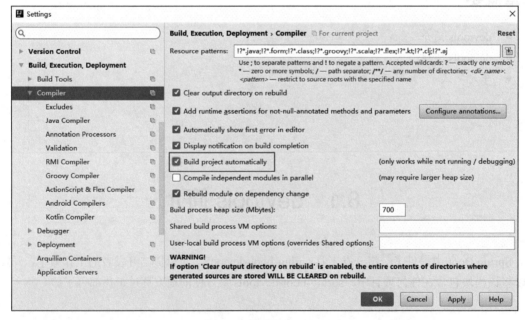

图 8-1

步骤02 按 Ctrl+Shift+Alt+/快捷键调出 Maintenance 页面，如图 8-2 所示。

图 8-2

单击 Registry，在新打开的 Registry 页面中，勾选 compiler.automake.allow.when.app.running 复选框，如图 8-3 所示。

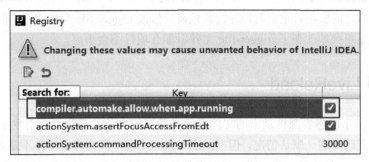

图 8-3

做完这两步配置之后，若开发者再次在 IntelliJ IDEA 中修改代码，则项目会自动重启。

8.2.2 基本原理

Spring Boot 中使用的自动重启技术涉及两个类加载器，一个是 baseclassloader，用来加载不会变化的类，例如项目引用的第三方的 jar；另一个是 restartclassloader，用来加载开发者自己写的会变化的类。当项目需要重启时，restartclassloader 将被一个新创建的类加载器代替，而 baseclassloader 则继续使用原来的，这种启动方式要比冷启动快很多，因为 baseclassloader 已经存在并且已经加载好。

8.2.3 自定义监控资源

默认情况下，/META-INF/maven、/META-INF/resources、resources、/static、/public 以及/templates 位置下资源的变化并不会触发重启，如果开发者想要对这些位置进行重定义，在 application.properties 中添加如下配置即可：

```
1  spring.devtools.restart.exclude=static/**
```

这表示从默认的不触发重启的目录中除去 static 目录，即 classpath:static 目录下的资源发生变化时也会导致项目重启。用户也可以反向配置需要监控的目录，配置方式如下：

```
1  spring.devtools.restart.additional-paths=src/main/resources/static
```

这个配置表示当 src/main/resources/static 目录下的文件发生变化时，自动重启项目。

由于项目的编码过程是一个连续的过程，并不是每修改一行代码就要重启项目，这样不仅浪费电脑性能，而且没有实际意义。鉴于这种情况，开发者也可以考虑使用触发文件，触发文件是一个特殊的文件，当这个文件发生变化时项目就会重启，配置方式如下：

```
spring.devtools.restart.trigger-file=.trigger-file
```

在项目 resources 目录下创建一个名为 .trigger-file 的文件,此时当开发者修改代码时,默认情况下项目不会重启,需要项目重启时,开发者只需要修改 .trigger-file 文件即可。但是注意,如果项目没有改变,只是单纯地改变了 .trigger-file 文件,那么项目不会重启。

8.2.4 使用 LiveReload

上一小节介绍了静态资源目录下的文件变化以及模板文件的变化不会引发重启,虽然开发者可以通过修改配置改变这一默认情况,但实际上并没有必要,因为静态文件不是 class。devtools 默认嵌入了 LiveReload 服务器,可以解决静态文件的热部署,LiveReload 可以在资源发生变化时自动触发浏览器更新,LiveReload 支持 Chrome、Firefox 以及 Safari。以 Chrome 为例,在 Chrome 应用商店搜索 LiveReload,结果如图 8-4 所示。

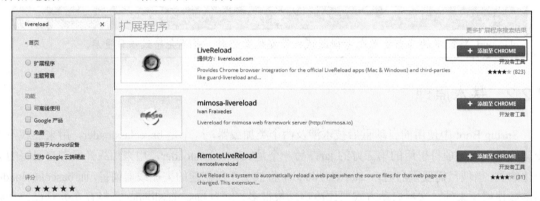

图 8-4

将第一个搜索结果添加到 Chrome 中,添加成功后,在 Chrome 右上角有一个 LiveReload 图标,如图 8-5 所示。

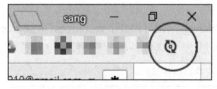

图 8-5

在浏览器中打开项目的页面,然后单击浏览器右上角的 LiveReload 按钮,开启 LiveReload 连接,此时当静态资源发生改变时,浏览器就会自动加载。如果开发者不想使用这一特性,可通过如下配置关闭:

```
spring.devtools.livereload.enabled=false
```

> **注 意**
>
> 建议开发者使用 LiveReload 策略而不是项目重启策略来实现静态资源的动态加载,因为项目重启所耗费的时间一般要超过 LiveReload。

> **提示**
>
> 在 Firefox 中安装 LiveReload，首先打开附加组件页面，如图 8-6 所示。然后在附加组件中搜索 LiveReload 并安装，用法和 Chrome 一致，这里不再赘述。

图 8-6

8.2.5 禁用自动重启

如果开发者添加了 spring-boot-devtools 依赖但是不想使用自动重启特性，那么可以关闭自动重启，代码如下：

```
spring.devtools.restart.enabled=false
```

也可以在 Java 代码中配置禁止自动重启，配置方式如下：

```
@Spring BootApplication
public class DevtoolsApplication {
    public static void main(String[] args) {
        System.setProperty("spring.devtools.restart.enabled", "false");
        SpringApplication.run(DevtoolsApplication.class, args);
    }
}
```

8.2.6 全局配置

如果项目模块众多，开发者可以在当前用户目录下创建 .spring-boot-devtools.properties 文件来对 devtools 进行全局配置，这个配置文件适用于当前计算机上任何使用了 devtools 模块的 Spring Boot 项目。以笔者的电脑为例，在 C:\Users\sang 目录下创建.spring-boot-devtools.properties 文件，内容如下：

```
1  spring.devtools.restart.trigger-file=.trigger-file
```

此时，就实现了使用触发文件触发项目重启。

8.3 单元测试

在本书前面的章节中，遇到需要测试的地方都是创建一个 Controller 进行测试，这样操作臃肿，效率低下，在 Spring Boot 中使用单元测试可以实现对每一个环节的代码进行测试。Spring Boot 中的单元测试与 Spring 中的测试一脉相承，但是又做了大量的简化，只需要少量的代码就能搭建一个测试环境，进而实现对 Controller、Service 或者 Dao 层的代码进行测试。接下来向读者介绍 Spring Boot 中单元测试的主要用法。

8.3.1 基本用法

当开发者使用 IntelliJ IDEA 或者在线创建一个 Spring Boot 项目时，创建成功后，默认都添加了 spring-boot-starter-test 依赖，并且创建好了测试类，代码如下：

```
1  @RunWith(SpringRunner.class)
2  @Spring BootTest
3  public class Test01ApplicationTests {
4      @Test
5      public void contextLoads() {
6      }
7  }
```

代码解释：

- 这里首先使用了@RunWith 注解，该注解将 JUnit 执行类修改为 SpringRunner，而 SpringRunner 是 Spring Framework 中测试类 SpringJUnit4ClassRunner 的别名。
- @Spring BootTest 注解除了提供 Spring TestContext 中的常规测试功能之外，还提供了其他特性：提供默认的 ContextLoader、自动搜索@Spring BootConfiguration、自定义环境属性、为不同的 webEnvironment 模式提供支持，这里的 webEnvironment 模式主要有 4 种：
  - MOCK，这种模式是当 classpath 下存在 servletAPIS 时，就会创建 WebApplicationContext 并提供一个 mockservlet 环境；当 classpath 下存在 Spring WebFlux 时，则创建 ReactiveWebApplicationContext；若都不存在，则创建一个常规的 ApplicationContext。
  - RANDOM_PORT，这种模式将提供一个真实的 Servlet 环境，使用内嵌的容器，但是端口随机。
  - DEFINED_PORT，这种模式也将提供一个真实的 Servlet 环境，使用内嵌的容器，但是使用定义好的端口。

> NONE，这种模式则加载一个普通的 ApplicationContext，不提供任何 Servlet 环境。这种一般不适用于 Web 测试。

- 在 Spring 测试中，开发者一般使用 @ContextConfiguration(classes =) 或者 @ContextConfiguration(locations =)来指定要加载的 Spring 配置，而在 Spring Boot 中则不需要这么麻烦，Spring Boot 中的 @*Test 注解将会去包含测试类的包下查找带有@SpringBootApplication 或者@Spring BootConfiguration 注解的主配置类。
- @Test 注解则来则 junit，junit 中的@After、@AfterClass、@Before、@BeforeClass、@Ignore 等注解一样可以在这里使用。

8.3.2 Service 测试

Service 层的测试就是常规测试，非常容易，例如现在有一个 HelloService 如下：

```
1  @Service
2  public class HelloService {
3      public String sayHello(String name) {
4          return "Hello " + name + " !";
5      }
6  }
```

需要对 HelloService 进行测试，直接在测试类中注入 HelloService 即可：

```
1  @RunWith(SpringRunner.class)
2  @Spring BootTest
3  public class Test01ApplicationTests {
4      @Autowired
5      HelloService helloService;
6      @Test
7      public void contextLoads() {
8          String hello = helloService.sayHello("Michael");
9          Assert.assertThat(hello,Matchers.is("Hello Michael !"));
10     }
11 }
```

在测试类中注入 HelloService，然后调用相关方法即可。第 9 行使用 Assert 判断测试结果是否正确。

8.3.3 Controller 测试

Controller 测试则要使用到 Mock 测试，即对一些不易获取的对象采用虚拟的对象来创建进而方便测试。而 Spring 中提供的 MockMvc 则提供了对 HTTP 请求的模拟，使开发者能够在不依赖网络环境的情况下实现对 Controller 的快速测试。例如有如下 Controller：

```
1  @RestController
2  public class HelloController {
```

```
3       @GetMapping("/hello")
4       public String hello(String name) {
5           return "Hello " + name + " !";
6       }
7       @PostMapping("/book")
8       public String addBook(@RequestBody Book book) {
9           return book.toString();
10      }
11  }
```

Controller 中涉及的实体类 Book 如下：

```
1   public class Book {
2       private Integer id;
3       private String name;
4       private String author;
5       //省略 getter/setter 方法
6   }
```

如果要对这个 Controller 进行测试，就需要借助 MockMvc，代码如下：

```
1   @RunWith(SpringRunner.class)
2   @SpringBootTest
3   public class Test01ApplicationTests {
4       @Autowired
5       HelloService helloService;
6       @Autowired
7       WebApplicationContext wac;
8       MockMvc mockMvc;
9       @Before
10      public void before() {
11          mockMvc = MockMvcBuilders.webAppContextSetup(wac).build();
12      }
13      @Test
14      public void test1() throws Exception {
15          MvcResult mvcResult = mockMvc.perform(
16                  MockMvcRequestBuilders
17                          .get("/hello")
18                          .contentType(MediaType.APPLICATION_FORM_URLENCODED)
19                          .param("name", "Michael"))
20                  .andExpect(MockMvcResultMatchers.status().isOk())
21                  .andDo(MockMvcResultHandlers.print())
22                  .andReturn();
23          System.out.println(mvcResult.getResponse().getContentAsString());
24      }
25      @Test
26      public void test2() throws Exception {
27          ObjectMapper om = new ObjectMapper();
28          Book book = new Book();
29          book.setAuthor("罗贯中");
30          book.setName("三国演义");
```

```
31              book.setId(1);
32              String s = om.writeValueAsString(book);
33              MvcResult mvcResult = mockMvc
34                      .perform(MockMvcRequestBuilders
35                          .post("/book")
36                          .contentType(MediaType.APPLICATION_JSON)
37                          .content(s))
38                      .andExpect(MockMvcResultMatchers.status().isOk())
39                      .andReturn();
40              System.out.println(mvcResult.getResponse().getContentAsString());
41          }
42      }
```

代码解释：

- 第 6、7 行注入一个 WebApplicationContext 用来模拟 ServletContext 环境。
- 第 8 行声明一个 MockMvc 对象，并在每个测试方法执行前进行 MockMvc 的初始化操作（第 9~12 行）。
- 第 15 行调用 MockMvc 中的 perform 方法开启一个 RequestBuilder 请求，具体的请求则通过 MockMvcRequestBuilders 进行构建，调用 MockMvcRequestBuilders 中的 get 方法表示发起一个 GET 请求，调用 post 方法则发起一个 POST 请求，其他的 DELETE 和 PUT 请求也是一样的，最后通过调用 param 方法设置请求参数。
- 第 20 行表示添加返回值的验证规则，利用 MockMvcResultMatchers 进行验证，这里表示验证响应码是否为 200。
- 第 21 行表示将请求详细信息打印到控制台。
- 第 22 行表示返回相应的 MvcResult，并在 23 行将之获取并打印出来。
- test2 方法演示了 POST 请求如何传递 JSON 数据，首先在 32 行将一个 book 对象转为一段 JSON，然后在 36 行设置请求的 contentType 为 APPLICATION-JSON，最后在 37 行设置 content 为上传的 JSON 即可。

除了 MockMvc 这种测试方式之外，Spring Boot 还专门提供了 TestRestTemplate 用来实现集成测试，若开发者使用了 @Spring BootTest 注解，则 TestRestTemplate 将自动可用，直接在测试类中注入即可。注意，如果要使用 TestRestTemplate 进行测试，需要将 @Spring BootTest 注解中 webEnvironment 属性的默认值由 WebEnvironment.MOCK 修改为 WebEnvironment.DEFINED_PORT 或者 WebEnvironment.RANDOM_PORT，因为这两种都是使用一个真实的 Servlet 环境而不是模拟的 Servlet 环境。其代码如下：

```
1   @RunWith(SpringRunner.class)
2   @Spring BootTest(webEnvironment = Spring BootTest.WebEnvironment.DEFINED_PORT)
3   public class Test01ApplicationTests {
4       @Autowired
5       TestRestTemplate restTemplate;
6       @Test
7       public void test3() {
8           ResponseEntity<String> hello = restTemplate.getForEntity("/hello?name={0}",
9   String.class, "Michael");
```

```
10            System.out.println(hello.getBody());
11      }
12 }
```

8.3.4 JSON 测试

开发者可以使用@JsonTest 测试 JSON 序列化和反序列化是否工作正常，该注解将自动配置 Jackson ObjectMapper、@JsonComponent 以及 Jackson Modules。如果开发者使用 Gson 代替 Jackson，该注解将配置 Gson。具体用法如下：

```
1  @RunWith(SpringRunner.class)
2  @JsonTest
3  public class JSONTest {
4      @Autowired
5      JacksonTester<Book> jacksonTester;
6      @Test
7      public void testSerialize() throws IOException {
8          Book book = new Book();
9          book.setId(1);
10         book.setName("三国演义");
11         book.setAuthor("罗贯中");
12         Assertions.assertThat(jacksonTester.write(book))
13                 .isEqualToJson("book.json");
14         Assertions.assertThat(jacksonTester.write(book))
15                 .hasJsonPathStringValue("@.name");
16         Assertions.assertThat(jacksonTester.write(book))
17                 .extractingJsonPathStringValue("@.name")
18                 .isEqualTo("三国演义");
19     }
20     @Test
21     public void testDeserialize() throws Exception {
22         String content = "{\"id\":1,\"name\":\"三国演义\",\"author\":\"罗贯中\"}";
23         Assertions.assertThat(jacksonTester.parseObject(content).getName())
24                 .isEqualTo("三国演义");
25     }
26 }
```

代码解释：

- 首先第 2 行添加 JacksonTester 注解，第 5 行注入 JacksonTester 进行 JSON 的序列化和反序列化测试。
- 第 12、13 行在序列化完成后判断序列化结果是否是所期待的 json，book.json 是一个定义在当前包下的 JSON 文件。
- 第 14、15 行判断对象序列化之后生成的 JSON 中是否有一个名为 name 的 key。
- 第 16~18 行判断序列化后 name 对应的值是否为"三国演义"。
- 第 23、24 行是反序列化，反序列化完成时判断对象的 name 属性值是否为"三国演义"。

book.json 文件如图 8-7 所示。

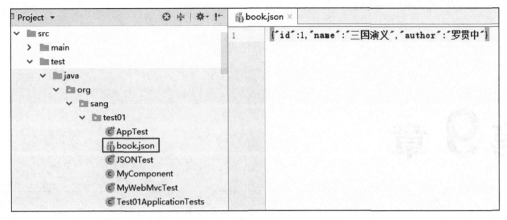

图 8-7

8.4 小　　结

本章向读者介绍了 Spring Boot 中的开发者工具和单元测试，开发者工具的一个核心功能就是热部署，结合 LiveReload 可以极大地缩短开发者等待编译的时间，有效提高开发效率；单元测试则与 Spring 单元测试一脉相承，但是又增加了许多功能，同时简化了测试代码，使开发者极大地节省了测试的编码时间。本章对于单元测试只是介绍了一些常用功能，如果读者想了解完整的单元测试功能，可以参考 Spring Boot 官方文档单元测试一节。

第 9 章

Spring Boot 缓存

本章概要

- Ehcache 2.x 缓存
- Redis 单机缓存
- Redis 集群缓存

Spring 3.1 中开始对缓存提供支持，核心思路是对方法的缓存，当开发者调用一个方法时，将方法的参数和返回值作为 key/value 缓存起来，当再次调用该方法时，如果缓存中有数据，就直接从缓存中获取，否则再去执行该方法。但是，Spring 中并未提供缓存的实现，而是提供了一套缓存 API，开发者可以自由选择缓存的实现，目前 Spring Boot 支持的缓存有如下几种：

- JCache (JSR-107)
- EhCache 2.x
- Hazelcast
- Infinispan
- Couchbase
- Redis
- Caffeine
- Simple

本章将介绍目前常用的缓存实现 Ehcache 2.x 和 Redis。由于 Spring 早已将缓存领域统一，因此无论使用哪种缓存实现，不同的只是缓存配置，开发者使用的缓存注解是一致的（Spring 缓存注解和各种缓存实现的关系就像 JDBC 和各种数据库驱动的关系一样）。

9.1　Ehcache 2.x 缓存

Ehcache 缓存在 Java 开发领域已是久负盛名，在 Spring Boot 中，只需要一个配置文件就可以将 Ehcache 集成到项目中。Ehcache 2.x 的使用步骤如下。

1. 创建项目，添加缓存依赖

创建 Spring Boot 项目，添加 spring-boot-starter-cache 依赖以及 Ehcache 依赖，代码如下：

```
1   <dependency>
2       <groupId>org.springframework.boot</groupId>
3       <artifactId>spring-boot-starter-cache</artifactId>
4   </dependency>
5   <dependency>
6       <groupId>net.sf.ehcache</groupId>
7       <artifactId>ehcache</artifactId>
8   </dependency>
9   <dependency>
10      <groupId>org.springframework.boot</groupId>
11      <artifactId>spring-boot-starter-web</artifactId>
12  </dependency>
```

2. 添加缓存配置文件

如果 Ehcache 的依赖存在，并且在 classpath 下有一个名为 ehcache.xml 的 Ehcache 配置文件，那么 EhCacheCacheManager 将会自动作为缓存的实现。因此，在 resources 目录下创建 ehcache.xml 文件作为 Ehcache 缓存的配置文件，代码如下：

```
1   <ehcache>
2   <diskStore path="java.io.tmpdir/cache"/>
3   <defaultCache
4           maxElementsInMemory="10000"
5           eternal="false"
6           timeToIdleSeconds="120"
7           timeToLiveSeconds="120"
8           overflowToDisk="false"
9           diskPersistent="false"
10          diskExpiryThreadIntervalSeconds="120"
11      />
12  <cache name="book_cache"
13          maxElementsInMemory="10000"
14          eternal="true"
15          timeToIdleSeconds="120"
16          timeToLiveSeconds="120"
17          overflowToDisk="true"
18          diskPersistent="true"
19          diskExpiryThreadIntervalSeconds="600"/>
```

```
20    </ehcache>
```

这是一个常规的 Ehcache 配置文件，提供了两个缓存策略，一个是默认的，另一个名为 book_cache。其中，name 表示缓存名称；maxElementsInMemory 表示缓存最大个数；eternal 表示缓存对象是否永久有效，一旦设置了永久有效，timeout 将不起作用；timeToIdleSeconds 表示缓存对象在失效前的允许闲置时间（单位：秒），当 eternal=false 对象不是永久有效时，该属性才生效；timeToLiveSeconds 表示缓存对象在失效前允许存活的时间（单位：秒），当 eternal=false 对象不是永久有效时，该属性才生效；overflowToDisk 表示当内存中的对象数量达到 maxElementsInMemory 时，Ehcache 是否将对象写到磁盘中；diskExpiryThreadIntervalSeconds 表示磁盘失效线程运行时间间隔。还有其他更为详细的 Ehcache 配置，这里就不一一介绍了。另外，如果开发者想自定义 Ehcache 配置文件的名称和位置，可以在 application.properties 中添加如下配置：

```
1    spring.cache.ehcache.config=classpath:config/another-config.xml
```

3. 开启缓存

在项目的入口类上添加@EnableCaching 注解开启缓存，代码如下：

```
1    @Spring BootApplication
2    @EnableCaching
3    public class CacheApplication {
4        public static void main(String[] args) {
5            SpringApplication.run(CacheApplication.class, args);
6        }
7    }
```

4. 创建 BookDao

创建 Book 实体类和 BookService，代码如下：

```
1    @Repository
2    @CacheConfig(cacheNames = "book_cache")
3    public class BookDao {
4        @Cacheable
5        public Book getBookById(Integer id) {
6            System.out.println("getBookById");
7            Book book = new Book();
8            book.setId(id);
9            book.setName("三国演义");
10           book.setAuthor("罗贯中");
11           return book;
12       }
13       @CachePut(key = "#book.id")
14       public Book updateBookById(Book book) {
15           System.out.println("updateBookById");
16           book.setName("三国演义2");
17           return book;
18       }
19       @CacheEvict(key = "#id")
20       public void deleteBookById(Integer id) {
21           System.out.println("deleteBookById");
```

```
22          }
23      }
24  public class Book implements Serializable {
25      private Integer id;
26      private String name;
27      private String author;
28      //省略getter/setter
29  }
```

代码解释：

- 在 BookDao 上添加@CacheConfig 注解指明使用的缓存的名字，这个配置可选，若不使用@CacheConfig 注解，则直接在@Cacheable 注解中指明缓存名字。
- 第 4 行在 getBookById 方法上添加@Cacheable 注解表示对该方法进行缓存，默认情况下，缓存的 key 是方法的参数，缓存的 value 是方法的返回值。当开发者在其他类中调用该方法时，首先会根据调用参数查看缓存中是否有相关数据，若有，则直接使用缓存数据，该方法不会执行，否则执行该方法，执行成功后将返回值缓存起来，但若是在当前类中调用该方法，则缓存不会生效。
- @Cacheable 注解中还有一个属性 condition 用来描述缓存的执行时机，例如@Cacheable(condition = "#id%2==0")表示当 id 对 2 取模为 0 时才进行缓存，否则不缓存。
- 如果开发者不想使用默认的 key，也可以像第 13 行和第 19 行一样自定义 key，第 13 行表示缓存的 key 为参数 book 对象中 id 的值，第 19 行表示缓存的 key 为参数 id。除了这种使用参数定义 key 的方式之外，Spring 还提供了一个 root 对象用来生成 key，如表 9-1 所示。

表 9-1　使用 root 对象生成 key

属性名称	属性描述	用法示例
methodName	当前方法名	#root.methodName
method	当前方法对象	#root.method.name
caches	当前方法使用的缓存	#root.caches[0].name
target	当前被调用的对象	#root.target
targetClass	当前被调用的对象的 class	#root.targetClass
args	当前方法参数数组	#root.args[0]

- 如果这些 key 不能够满足开发需求，开发者也可以自定义缓存 key 的生成器 KeyGenerator，代码如下：

```
1   @Component
2   public class MyKeyGenerator implements KeyGenerator {
3       @Override
4       public Object generate(Object target, Method method, Object... params) {
5           return Arrays.toString(params);
6       }
7   }
8   @Service
9   @CacheConfig(cacheNames = "book_cache")
10  public class BookDao {
```

```
11      @Autowired
12      MyKeyGenerator myKeyGenerator;
13      @Cacheable(keyGenerator = "myKeyGenerator")
14      public Book getBookById(Integer id) {
15          System.out.println("getBookById");
16      …
17      …
18          return book;
19      }
20  }
```

自定义 MyKeyGenerator 实现 KeyGenerator 接口，然后实现该接口中的 generate 方法，该方法的三个参数分别是当前对象、当前请求的方法以及方法的参数，开发者可根据这些信息组成一个新的 key 返回，返回值就是缓存的 key。第 13 行在@Cacheable 注解中引用 MyKeyGenerator 实例即可。

- 第 13 行@CachePut 注解一般用于数据更新方法上，与@Cacheable 注解不同，添加了@CachePut 注解的方法每次在执行时都不去检查缓存中是否有数据，而是直接执行方法，然后将方法的执行结果缓存起来，如果该 key 对应的数据已经被缓存起来了，就会覆盖之前的数据，这样可以避免再次加载数据时获取到脏数据。同时，@CachePut 具有和@Cacheable 类似的属性，这里不再赘述。
- 第 19 行@CacheEvict 注解一般用于删除方法上，表示移除一个 key 对应的缓存。@CacheEvict 注解有两个特殊的属性：allEntries 和 beforeInvocation，其中 allEntries 表示是否将所有的缓存数据都移除，默认为 false，beforeInvocation 表示是否在方法执行之前移除缓存中的数据，默认为 false，即在方法执行之后移除缓存中的数据。

5. 创建测试类

创建测试类，对 Service 中的方法进行测试，代码如下：

```
1   @RunWith(SpringRunner.class)
2   @Spring BootTest
3   public class CacheApplicationTests {
4       @Autowired
5       BookDao bookDao;
6       @Test
7       public void contextLoads() {
8           bookDao.getBookById(1);
9           bookDao.getBookById(1);
10          bookDao.deleteBookById(1);
11          Book b3 = bookDao.getBookById(1);
12          System.out.println("b3:"+b3);
13          Book b = new Book();
14          b.setName("三国演义");
15          b.setAuthor("罗贯中");
16          b.setId(1);
17          bookDao.updateBookById(b);
18          Book b4 = bookDao.getBookById(1);
19          System.out.println("b4:"+b4);
```

```
20      }
21 }
```

执行该方法，控制台打印日志如图 9-1 所示。

```
2018-07-25 21:37:28.494  INFO 7588 --- [
getBookById
deleteBookById
getBookById
b3:Book{id=1, name='三国演义', author='罗贯中'}
updateBookById
b4:Book{id=1, name='三国演义2', author='罗贯中'}
```

图 9-1

一开始执行了两个查询，但是查询方法只打印了一次，因为第二次使用了缓存。接下来执行了删除方法，删除方法执行完之后再次执行查询，查询方法又被执行了，因为在删除方法中缓存已经被删除了。再接下来执行更新方法，更新方法中不仅更新数据，也更新了缓存，所以在最后的查询方法中，查询方法日志没打印，说明该方法没执行，而是使用了缓存中的数据，而缓存中的数据已经被更新了。

9.2　Redis 单机缓存

和 Ehcache 一样，如果在 classpath 下存在 Redis 并且 Redis 已经配置好了，此时默认就会使用 RedisCacheManager 作为缓存提供者。Redis 单机使用步骤如下。

1．创建项目，添加缓存依赖

创建 Spring Boot 项目，添加 spring-boot-starter-cache 和 Redis 依赖，代码如下：

```
1   <dependency>
2       <groupId>org.springframework.boot</groupId>
3       <artifactId>spring-boot-starter-cache</artifactId>
4   </dependency>
5   <dependency>
6       <groupId>org.springframework.boot</groupId>
7       <artifactId>spring-boot-starter-web</artifactId>
8   </dependency>
9   <dependency>
10      <groupId>org.springframework.boot</groupId>
11      <artifactId>spring-boot-starter-data-redis</artifactId>
12      <exclusions>
13          <exclusion>
14              <groupId>io.lettuce</groupId>
15              <artifactId>lettuce-core</artifactId>
16          </exclusion>
```

```
17         </exclusions>
18     </dependency>
19     <dependency>
20         <groupId>redis.clients</groupId>
21         <artifactId>jedis</artifactId>
22     </dependency>
```

2. 缓存配置

Redis 单机缓存只需要开发者在 application.properties 中进行 Redis 配置及缓存配置即可，代码如下：

```
1  #缓存配置
2  spring.cache.cache-names=c1,c2
3  spring.cache.redis.time-to-live=1800s
4  #Redis 配置
5  spring.redis.database=0
6  spring.redis.host=192.168.66.129
7  spring.redis.port=6379
8  spring.redis.password=123@456
9  spring.redis.jedis.pool.max-active=8
10 spring.redis.jedis.pool.max-idle=8
11 spring.redis.jedis.pool.max-wait=-1ms
12 spring.redis.jedis.pool.min-idle=0
```

代码解释：

- 第 2 行是配置缓存名称。Redis 中的 key 都有一个前缀，默认前缀就是 "缓存名::"。
- 第 3 行配置缓存有效期，即 Redis 中 key 的过期时间。
- 第 5~12 行是 Redis 配置，具体含义可以参考本书 6.1 节。

3. 开启缓存

接下来在项目入口类中开启缓存，代码如下：

```
1  @SpringBootApplication
2  @EnableCaching
3  public class RediscacheApplication {
4      public static void main(String[] args) {
5          SpringApplication.run(RediscacheApplication.class, args);
6      }
7  }
```

最后创建 BookDao 和测试的步骤与 9.1 节的第 4、5 步一致，这里就不再赘述。

9.3 Redis 集群缓存

不同于 Redis 单机缓存，Redis 集群缓存的配置要复杂一些，主要体现在配置上，缓存的使用还是和 9.1 节中介绍的一样。搭建 Redis 集群缓存主要分为三个步骤：①搭建 Redis 集群；②配置缓存；③使用缓存。下面按照这三个步骤向读者介绍 Redis 集群缓存的搭建过程。

9.3.1 搭建 Redis 集群

Redis 集群的搭建过程在 6.1.4 小节已经介绍过了，本案例中采用的 Redis 集群案例和 6.1.4 小节搭建的 Redis 集群一致，都是 8 台 Redis 实例，4 主 4 从，端口从 8001 到 8008，具体搭建过程这里就不赘述了，读者可以参考 6.1.4 小节。Redis 集群搭建成功后，通过 Spring Data Redis 连接 Redis 集群，这一段配置也和 6.1.4 小节中的一致，因此这里也不赘述。总之，读者需要参考 6.1.4 小节先将 Redis 集群搭建成功，并且能够在 Spring Boot 中通过 RedisTemplate 访问成功。

9.3.2 配置缓存

当 Redis 集群搭建成功，并且能够从 Spring Boot 项目中访问 Redis 集群后，只需要进行简单的 Redis 缓存配置即可，代码如下：

```
@Configuration
public class RedisCacheConfig {
    @Autowired
    RedisConnectionFactory conFactory;
    @Bean
    RedisCacheManager redisCacheManager() {
        Map<String, RedisCacheConfiguration> configMap = new HashMap<>();
        RedisCacheConfiguration redisCacheConfig =
                RedisCacheConfiguration.defaultCacheConfig()
                        .prefixKeysWith("sang:")
.disableCachingNullValues()
                        .entryTtl(Duration.ofMinutes(30));
        configMap.put("c1", redisCacheConfig);
        RedisCacheWriter cacheWriter =
                RedisCacheWriter.nonLockingRedisCacheWriter(conFactory);
        RedisCacheManager redisCacheManager =
                new RedisCacheManager(
                        cacheWriter,
                        RedisCacheConfiguration.defaultCacheConfig(),
                        configMap);
        return redisCacheManager;
    }
}
```

代码解释：

- 在配置 Redis 集群时，已经向 Spring 容器中注册了一个 JedisConnectionFactory 的实例，这里将之注入到 RedisCacheConfig 配置文件中备用（RedisConnectionFactory 是 JedisConnectionFactory 的父类）。

- 在 RedisCacheConfig 中提供 RedisCacheManager 的实例，该实例的构建需要三个参数，第一个参数是一个 cacheWriter，直接通过 nonLockingRedisCacheWriter 方法构造出来即可；第二个参数是默认的缓存配置；第三个参数是提前定义好的缓存配置。
- RedisCacheManager 构造方法中第三个参数是一个提前定义好的缓存参数，它是一个 Map 类型的参数，该 Map 中的 key 就是指缓存名字，value 就是该名称的缓存所对应的缓存配置，例如 key 的前缀、缓存过期时间等，若缓存注解中使用的缓存名称不存在于 Map 中，则使用 RedisCacheManager 构造方法中第二个参数所定义的缓存策略进行数据缓存。例如如下两个缓存配置：

```
1  @Cacheable(value = "c1")
2  @Cacheable(value = "c2")
```

第 1 行的注解中，c1 存在于 configMap 集合中，因此使用的缓存策略是 configMap 集合中 c1 所对应的缓存策略，c2 不存在于 configMap 集合中，因此使用的缓存策略是默认的缓存策略。

- 本案例中默认缓存策略通过调用 RedisCacheConfiguration 中的 defaultCacheConfig 方法获取，该方法部分源码如下：

```
1  public static RedisCacheConfiguration defaultCacheConfig() {
2  …
3  …
4     return new RedisCacheConfiguration(Duration.ZERO, true, true,
5  CacheKeyPrefix.simple(),
6         SerializationPair.fromSerializer(new StringRedisSerializer()),
7         SerializationPair.fromSerializer(new JdkSerializationRedisSerializer()),
8  conversionService);
   }
```

由这一段源码可以看到，默认的缓存过期时间为 0，即永不过期；第二个参数 true 表示允许存储 null，第三个参数 true 表示开启 key 的前缀，第四个参数表示 key 的默认前缀是 "缓存名::"，接下来两个参数表示 key 和 value 的序列化方式，最后一个参数则是一个类型转换器。

- 本案例中第 8~12 行是一个自定义的缓存配置，第 10 行设置了 key 的前缀为 "sang:"，第 11 行禁止缓存一个 null，第 12 行设置缓存的过期时间为 30 分钟。

9.3.3 使用缓存

缓存配置完成后，接下来首先在项目启动类中通过@EnableCaching 注解开启缓存，代码如下：

```
1  @Spring BootApplication
2  @EnableCaching
3  public class RedisclustercacheApplication {
4     public static void main(String[] args) {
5        SpringApplication.run(RedisclustercacheApplication.class, args);
6     }
7  }
```

然后创建一个 BookDao 使用缓存，代码如下：

```
@Repository
public class BookDao {
    @Cacheable(value = "c1")
    public String getBookById(Integer id) {
        System.out.println("getBookById");
        return "这本书是三国演义";
    }
    @CachePut(value = "c1")
    public String updateBookById(Integer id) {
        return "这是全新的三国演义";
    }
    @CacheEvict(value = "c1")
    public void deleteById(Integer id) {
        System.out.println("deleteById");
    }
    @Cacheable(value = "c2")
    public String getBookById2(Integer id) {
        System.out.println("getBookById2");
        return "这本书是红楼梦";
    }
}
```

最后创建单元测试，代码如下：

```
@RunWith(SpringRunner.class)
@SpringBootTest
public class RedisclustercacheApplicationTests {
    @Autowired
    BookDao bookDao;
    @Test
    public void contextLoads() {
        bookDao.getBookById(100);
        String book = bookDao.getBookById(100);
        System.out.println(book);
        bookDao.updateBookById(100);
        String book2 = bookDao.getBookById(100);
        System.out.println(book2);
        bookDao.deleteById(100);
        bookDao.getBookById(100);
        bookDao.getBookById2(99);

    }
}
```

单元测试运行结果如图 9-2 所示。

```
2018-07-26 22:55:04.413  INFO 4004 --- [
getBookById
这本书是三国演义
这是全新的三国演义
deleteById
getBookById
getBookById2
2018-07-26 22:55:04.737  INFO 4004 --- [
```

图 9-2

由单元测试可以看到，一开始做了两次查询，但是查询方法只调用了一次，因为第二次使用了缓存；接下来执行了更新，当更新成功后再去查询，此时缓存也已更新成功；接下来执行了删除，删除成功后再去执行查询，查询方法又被调用，说明缓存也已经被删除了；最后查询了一个 id 为 99 的记录，这次使用的是默认缓存配置。在 Redis 服务器上也可以看到缓存结果，如图 9-3 所示。

```
192.168.66.129:8002> get sang:100
"\xac\xed\x00\x05t\x00\x18\xe8\xbf\x99\xe6\x9c\xac\xe4\xb9
\xa6\xe6\x98\xaf\xe4\xb8\x89\xe5\x9b\xbd\xe6\xbc\x94\xe4\x
b9\x89"
192.168.66.129:8002> ttl sang:100
(integer) 1553
192.168.66.129:8002> get c2::99
-> Redirected to slot [1988] located at 192.168.66.129:800
1
"\xac\xed\x00\x05t\x00\x15\xe8\xbf\x99\xe6\x9c\xac\xe4\xb9
\xa6\xe6\x98\xaf\xe7\xba\xa2\xe6\xa5\xbc\xe6\xa2\xa6"
192.168.66.129:8001> ttl c2::99
(integer) -1
192.168.66.129:8001>
```

图 9-3

id 为 100 的记录使用的缓存名为 c1，因此 key 的前缀是 "sang:"，这是上文配置的，过期时间还剩 1553 秒（上文配置的过期时间是 30 分钟）；而 id 为 99 的记录使用的缓存名称为 c2，因此使用了默认的缓存配置，默认的前缀为 "缓存名::"，即 "c2::"，默认的过期时间是永不过期。

9.4 小　　结

本章向读者介绍了两种常见的缓存技术 Ehcache 和 Redis，其中 Redis 又分为单机缓存和集群缓存。Ehcache 部署简单，使用门槛较低，操作简便，但是功能较少，可扩展性较弱；Redis 则需要单独部署服务器，单机版的 Redis 缓存基本上做到了开箱即用，集群版的 Redis 缓存虽然配置烦琐，但是具有良好的扩展性与安全性，开发者在开发中可根据实际情况选择不同的缓存实现策略。

第 10 章

Spring Boot 安全管理

本章概要

- Spring Security 基本配置
- 基于数据库的认证
- 高级配置
- OAuth 2
- Spring Boot 整合 Shiro

安全可以说是公司的红线了，一般项目都会有严格的认证和授权操作，在 Java 开发领域常见的安全框架有 Shiro 和 Spring Security。Shiro 是一个轻量级的安全管理框架，提供了认证、授权、会话管理、密码管理、缓存管理等功能，Spring Security 是一个相对复杂的安全管理框架，功能比 Shiro 更加强大，权限控制细粒度更高，对 OAuth 2 的支持也更友好，又因为 Spring Security 源自 Spring 家族，因此可以和 Spring 框架无缝整合，特别是 Spring Boot 中提供的自动化配置方案，可以让 Spring Security 的使用更加便捷。本章将主要介绍 Spring Security 以及 Shiro 在 Spring Boot 中的使用。

10.1　Spring Security 的基本配置

Spring Boot 针对 Spring Security 提供了自动化配置方案，因此可以使 Spring Security 非常容易地整合进 Spring Boot 项目中，这也是在 Spring Boot 项目中使用 Spring Security 的优势。

10.1.1 基本用法

基本整合步骤如下。

1. 创建项目，添加依赖

创建一个 Spring Boot Web 项目，然后添加 spring-boot-starter-security 依赖即可，代码如下：

```
1  <dependency>
2      <groupId>org.springframework.boot</groupId>
3      <artifactId>spring-boot-starter-security</artifactId>
4  </dependency>
5  <dependency>
6      <groupId>org.springframework.boot</groupId>
7      <artifactId>spring-boot-starter-web</artifactId>
8  </dependency>
```

只要开发者在项目中添加了 spring-boot-starter-security 依赖，项目中所有资源都会被保护起来。

2. 添加 hello 接口

接下来在项目中添加一个简单的 /hello 接口，内容如下：

```
1  @RestController
2  public class HelloController {
3      @GetMapping("/hello")
4      public String hello() {
5          return "Hello";
6      }
7  }
```

3. 启动项目测试

接下来启动项目，启动成功后，访问 /hello 接口会自动跳转到登录页面，这个登录页面是由 Spring Security 提供的，如图 10-1 所示。

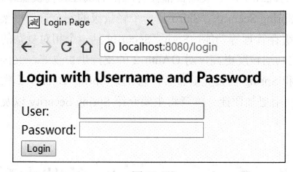

图 10-1

默认的用户名是 user，默认的登录密码则在每次启动项目时随机生成，查看项目启动日志，如图 10-2 所示。

```
2018-07-27 17:25:36.808  INFO 4528 --- [           main] .s.s.UserDetailsS

Using generated security password: 38000dff-a45a-4a6d-adb7-871216f19286

2018-07-27 17:25:36.980  INFO 4528 --- [           main] o.s.s.web.Defaul
```

图 10-2

从项目启动日志中可以看到默认的登录密码，登录成功后，用户就可以访问"/hello"接口了。

10.1.2 配置用户名和密码

如果开发者对默认的用户名和密码不满意，可以在 application.properties 中配置默认的用户名、密码以及用户角色，配置方式如下：

```
1  spring.security.user.name=sang
2  spring.security.user.password=123
3  spring.security.user.roles=admin
```

当开发者在 application.properties 中配置了默认的用户名和密码后，再次启动项目，项目启动日志就不会打印出随机生成的密码了，用户可直接使用配置好的用户名和密码登录，登录成功后，用户还具有一个角色——admin。

10.1.3 基于内存的认证

当然，开发者也可以自定义类继承自 WebSecurityConfigurerAdapter，进而实现对 Spring Security 更多的自定义配置，例如基于内存的认证，配置方式如下：

```
1   @Configuration
2   public class MyWebSecurityConfig extends WebSecurityConfigurerAdapter {
3       @Bean
4       PasswordEncoder passwordEncoder() {
5           return NoOpPasswordEncoder.getInstance();
6       }
7       @Override
8       protected void configure(AuthenticationManagerBuilder auth) throws Exception {
9           auth.inMemoryAuthentication()
10                  .withUser("admin").password("123").roles("ADMIN", "USER")
11                  .and()
12                  .withUser("sang").password("123").roles("USER");
13      }
14  }
```

代码解释：

- 自定义 MyWebSecurityConfig 继承自 WebSecurityConfigurerAdapter，并重写 configure(AuthenticationManagerBuilder auth)方法，在该方法中配置两个用户，一个用户名是

admin，密码 123，具备两个角色 ADMIN 和 USER；另一个用户名是 sang，密码是 123，具备一个角色 USER。
- 本案例使用的 Spring Security 版本是 5.0.6，在 Spring Security 5.x 中引入了多种密码加密方式，开发者必须指定一种，本案例使用 NoOpPasswordEncoder，即不对密码进行加密。

> **注 意**
> 基于内存的用户配置在配置角色时不需要添加 "ROLE_" 前缀，这点和 10.2 节中基于数据库的认证有差别。

配置完成后，重启 Spring Boot 项目，就可以使用这里配置的两个用户进行登录了。

10.1.4　HttpSecurity

虽然现在可以实现认证功能，但是受保护的资源都是默认的，而且也不能根据实际情况进行角色管理，如果要实现这些功能，就需要重写 WebSecurityConfigurerAdapter 中的另一个方法，代码如下：

```
1   @Configuration
2   public class MyWebSecurityConfig extends WebSecurityConfigurerAdapter {
3       @Bean
4       PasswordEncoder passwordEncoder() {
5           return NoOpPasswordEncoder.getInstance();
6       }
7       @Override
8       protected void configure(AuthenticationManagerBuilder auth) throws Exception {
9           auth.inMemoryAuthentication()
10                  .withUser("root").password("123").roles("ADMIN", "DBA")
11                  .and()
12                  .withUser("admin").password("123").roles("ADMIN", "USER")
13                  .and()
14                  .withUser("sang").password("123").roles("USER");
15      }
16      @Override
17      protected void configure(HttpSecurity http) throws Exception {
18          http.authorizeRequests()
19                  .antMatchers("/admin/**")
20                  .hasRole("ADMIN")
21                  .antMatchers("/user/**")
22                  .access("hasAnyRole('ADMIN','USER')")
23                  .antMatchers("/db/**")
24                  .access("hasRole('ADMIN') and hasRole('DBA')")
25                  .anyRequest()
26                  .authenticated()
27                  .and()
28                  .formLogin()
29                  .loginProcessingUrl("/login")
30                  .permitAll()
```

```
31              .and()
32              .csrf()
33              .disable();
34      }
35 }
```

代码解释：

- 首先配置了三个用户，root 用户具备 ADMIN 和 DBA 的角色，admin 用户具备 ADMIN 和 USER 的角色，sang 用户具备 USER 的角色。
- 第 18 行调用 authorizeRequests()方法开启 HttpSecurity 的配置，第 19~24 行配置分别表示用户访问 "/admin/**" 模式的 URL 必须具备 ADMIN 的角色；用户访问 "/user/**" 模式的 URL 必须具备 ADMIN 或 USER 的角色；用户访问 "/db/**"模式的 URL 必须具备 ADMIN 和 DBA 的角色。
- 第 25、26 行表示除了前面定义的 URL 模式之外，用户访问其他的 URL 都必须认证后访问（登录后访问）。
- 第 27~30 行表示开启表单登录，即读者一开始看到的登录页面，同时配置了登录接口为 "/login"，即可以直接调用 "/login" 接口，发起一个 POST 请求进行登录，登录参数中用户名必须命名为 username，密码必须命名为 password，配置 loginProcessingUrl 接口主要是方便 Ajax 或者移动端调用登录接口。最后还配置了 permitAll，表示和登录相关的接口都不需要认证即可访问。
- 第 32、33 行表示关闭 csrf。

配置完成后，接下来在 Controller 中添加如下接口进行测试：

```
1  @RestController
2  public class HelloController {
3      @GetMapping("/admin/hello")
4      public String admin() {
5          return "hello admin!";
6      }
7      @GetMapping("/user/hello")
8      public String user() {
9          return "hello user!";
10     }
11     @GetMapping("/db/hello")
12     public String dba() {
13         return "hello dba!";
14     }
15     @GetMapping("/hello")
16     public String hello() {
17         return "hello";
18     }
19 }
```

根据上文的配置，"/admin/hello" 接口 root 和 admin 用户具有访问权限；"/user/hello" 接口 admin 和 sang 用户具有访问权限；"/db/hello" 路径则只有 root 用户具有访问权限。浏览器中的测试比较容易，这里不再赘述。

10.1.5 登录表单详细配置

迄今为止，登录表单一直使用 Spring Security 提供的页面，登录成功后也是默认的页面跳转，但是，前后端分离正在成为企业级应用开发的主流，在前后端分离的开发方式中，前后端的数据交互通过 JSON 进行，这时，登录成功后就不是页面跳转了，而是一段 JSON 提示。要实现这些功能，只需要继续完善上文的配置，代码如下：

```
.and()
.formLogin()
.loginPage("/login_page")
.loginProcessingUrl("/login")
.usernameParameter("name")
.passwordParameter("passwd")
.successHandler(new AuthenticationSuccessHandler() {
    @Override
    public void onAuthenticationSuccess(HttpServletRequest req,
                                       HttpServletResponse resp,
                                       Authentication auth)
            throws IOException {
Object principal = auth.getPrincipal();
        resp.setContentType("application/json;charset=utf-8");
        PrintWriter out = resp.getWriter();
        resp.setStatus(200);
        Map<String, Object> map = new HashMap<>();
        map.put("status", 200);
        map.put("msg", principal);
        ObjectMapper om = new ObjectMapper();
        out.write(om.writeValueAsString(map));
        out.flush();
        out.close();
    }
})
.failureHandler(new AuthenticationFailureHandler() {
    @Override
    public void onAuthenticationFailure(HttpServletRequest req,
                                       HttpServletResponse resp,
                                       AuthenticationException e)
            throws IOException {
        resp.setContentType("application/json;charset=utf-8");
        PrintWriter out = resp.getWriter();
        resp.setStatus(401);
        Map<String, Object> map = new HashMap<>();
        map.put("status", 401);
        if (e instanceof LockedException) {
            map.put("msg", "账户被锁定，登录失败!");
        } else if (e instanceof BadCredentialsException) {
```

```
40                map.put("msg", "账户名或密码输入错误，登录失败!");
41            } else if (e instanceof DisabledException) {
42                map.put("msg", "账户被禁用，登录失败!");
43            } else if (e instanceof AccountExpiredException) {
44                map.put("msg", "账户已过期，登录失败!");
45            } else if (e instanceof CredentialsExpiredException) {
46                map.put("msg", "密码已过期，登录失败!");
47            }else{
48                map.put("msg", "登录失败!");
49            }
50            ObjectMapper om = new ObjectMapper();
51            out.write(om.writeValueAsString(map));
52            out.flush();
53            out.close();
54        }
55    })
56    .permitAll()
57    .and()
```

代码解释：

- 第 3 行配置了 loginPage，即登录页面，配置了 loginPage 后，如果用户未获授权就访问一个需要授权才能访问的接口，就会自动跳转到 login_page 页面让用户登录，这个 login_page 就是开发者自定义的登录页面，而不再是 Spring Security 提供的默认登录页。
- 第 4 行配置了 loginProcessingUrl，表示登录请求处理接口，无论是自定义登录页面还是移动端登录，都需要使用该接口。
- 第 5、6 行定义了认证所需的用户名和密码的参数名，默认用户名参数是 username，密码参数是 password，可以在这里自定义。
- 第 7~25 行定义了登录成功的处理逻辑。用户登录成功后可以跳转到某一个页面，也可以返回一段 JSON，这个要看具体业务逻辑，本案例假设是第二种，用户登录成功后，返回一段登录成功的 JSON。onAuthenticationSuccess 方法的第三个参数一般用来获取当前登录用户的信息，在登录成功后，可以获取当前登录用户的信息一起返回给客户端。
- 第 26~54 行定义了登录失败的处理逻辑，和登录成功类似，不同的是，登录失败的回调方法里有一个 AuthenticationException 参数，通过这个异常参数可以获取登录失败的原因，进而给用户一个明确的提示。

配置完成后，使用 Postman 进行登录测试，如图 10-3 所示。

登录请求参数用户名是 name，密码是 passwd，登录成功后返回用户的基本信息，密码已经过滤掉了。如果登录失败，也会有相应的提示，如图 10-4 所示。

图 10-3

图 10-4

10.1.6　注销登录配置

如果想要注销登录，也只需要提供简单的配置即可，代码如下：

```
.and()
.logout()
.logoutUrl("/logout")
.clearAuthentication(true)
.invalidateHttpSession(true)
.addLogoutHandler(new LogoutHandler() {
   @Override
   public void logout(HttpServletRequest req,
              HttpServletResponse resp,
              Authentication auth) {
   }
```

```
12      })
13      .logoutSuccessHandler(new LogoutSuccessHandler() {
14          @Override
15          public void onLogoutSuccess(HttpServletRequest req,
16                           HttpServletResponse resp,
17                           Authentication auth)
18              throws IOException {
19              resp.sendRedirect("/login_page");
20          }
21      })
22      .and()
```

代码解释：

- 第 2 行表示开启注销登录的配置。
- 第 3 行表示配置注销登录请求 URL 为 "/logout"，默认也是 "/logout"。
- 第 4 行表示是否清除身份认证信息，默认为 true，表示清除。
- 第 5 行表示是否使 Session 失效，默认为 true。
- 第 6 行配置一个 LogoutHandler，开发者可以在 LogoutHandler 中完成一些数据清除工作，例如 Cookie 的清除。Spring Security 提供了一些常见的实现，如图 10-5 所示。
- 第 13 行配置一个 LogoutSuccessHandler，开发者可以在这里处理注销成功后的业务逻辑，例如返回一段 JSON 提示或者跳转到登录页面等。

图 10-5

10.1.7 多个 HttpSecurity

如果业务比较复杂，开发者也可以配置多个 HttpSecurity，实现对 WebSecurityConfigurerAdapter 的多次扩展，代码如下：

```
1   @Configuration
2   public class MultiHttpSecurityConfig{
3       @Bean
4       PasswordEncoder passwordEncoder() {
5           return NoOpPasswordEncoder.getInstance();
6       }
7       @Autowired
8       protected void configure(AuthenticationManagerBuilder auth) throws Exception {
9           auth.inMemoryAuthentication()
```

```
10                .withUser("admin").password("123").roles("ADMIN", "USER")
11                .and()
12                .withUser("sang").password("123").roles("USER");
13     }
14     @Configuration
15     @Order(1)
16     public static class AdminSecurityConfig extends WebSecurityConfigurerAdapter{
17         @Override
18         protected void configure(HttpSecurity http) throws Exception {
19             http.antMatcher("/admin/**").authorizeRequests()
20                     .anyRequest().hasRole("ADMIN");
21         }
22     }
23     @Configuration
24     public static class OtherSecurityConfig extends WebSecurityConfigurerAdapter{
25         @Override
26         protected void configure(HttpSecurity http) throws Exception {
27             http.authorizeRequests()
28                     .anyRequest().authenticated()
29                     .and()
30                     .formLogin()
31                     .loginProcessingUrl("/login")
32                     .permitAll()
33                     .and()
34                     .csrf()
35                     .disable();
36         }
37     }
38 }
```

代码解释：

- 配置多个 HttpSecurity 时，MultiHttpSecurityConfig 不需要继承 WebSecurityConfigurerAdapter，在 MultiHttpSecurityConfig 中创建静态内部类继承 WebSecurityConfigurerAdapter 即可，静态内部类上添加@Configuration 注解和@Order 注解，@Order 注解表示该配置的优先级，数字越小优先级越大，未加@Order 注解的配置优先级最小。
- 第 14~22 行配置表示该类主要用来处理 "/admin/**" 模式的 URL，其他的 URL 将在第 23~37 行配置的 HttpSecurity 中进行处理。

10.1.8 密码加密

1. 为什么要加密

2011 年 12 月 21 日，有人在网络上公开了一个包含 600 万个 CSDN 用户资料的数据库，数据全部为明文存储，包含用户名、密码以及注册邮箱。事件发生后，CSDN 在微博、官方网站等渠道发出了声明，解释说此数据库是 2009 年备份所用的，因不明原因泄露，已经向警方报案，后又在官网发出了公开道歉信。在接下来的十多天里，金山、网易、京东、当当、新浪等多家公司被卷入这次事件中。整个事件中最触目惊心的莫过于 CSDN 把用户密码明文存储，由于很多用户是多个网站共用一个密码，因此一个网站密码泄露就会造成很大的安全隐患。有了这么多前车之鉴，我们现在做系统时，密码都要加密处理。

2. 加密方案

密码加密一般会用到散列函数，又称散列算法、哈希函数，这是一种从任何数据中创建数字"指纹"的方法。散列函数把消息或数据压缩成摘要，使得数据量变小，将数据的格式固定下来，然后将数据打乱混合，重新创建一个散列值。散列值通常用一个短的随机字母和数字组成的字符串来代表。好的散列函数在输入域中很少出现散列冲突。在散列表和数据处理中，不抑制冲突来区别数据会使得数据库记录更难找到。我们常用的散列函数有 MD5 消息摘要算法、安全散列算法（Secure Hash Algorithm）。

但是仅仅使用散列函数还不够，为了增加密码的安全性，一般在密码加密过程中还需要加盐，所谓的盐可以是一个随机数，也可以是用户名，加盐之后，即使密码明文相同的用户生成的密码，密文也不相同，这可以极大地提高密码的安全性。但是传统的加盐方式需要在数据库中有专门的字段来记录盐值，这个字段可能是用户名字段（因为用户名唯一），也可能是一个专门记录盐值的字段，这样的配置比较烦琐。Spring Security 提供了多种密码加密方案，官方推荐使用 BCryptPasswordEncoder，BCryptPasswordEncoder 使用 BCrypt 强哈希函数，开发者在使用时可以选择提供 strength 和 SecureRandom 实例。strength 越大，密钥的迭代次数越多，密钥迭代次数为 2^strength。strength 取值在 4~31 之间，默认为 10。

3. 实践

在 Spring Boot 中配置密码加密非常容易，只需要修改上文配置的 PasswordEncoder 这个 Bean 的实现即可，代码如下：

```
@Bean
PasswordEncoder passwordEncoder() {
    return new BCryptPasswordEncoder(10);
}
```

创建 BCryptPasswordEncoder 时传入的参数 10 就是 strength，即密钥的迭代次数（也可以不配置，默认为 10）。同时，配置的内存用户的密码也不再是 123 了，代码如下：

```
auth.inMemoryAuthentication()
    .withUser("admin")
    .password("$2a$10$RMuFXGQ5AtH4wOvkUqyvuecpqUSeoxZYqilXzbz50dceRsga.WYiq")
    .roles("ADMIN", "USER")
    .and()
    .withUser("sang")
    .password("$2a$10$eUHbAOMq4bpxTvOVz33LIehLe3fu6NwqC9tdOcxJXEhyZ4simqXTC")
    .roles("USER");
```

这里的密码就是使用 BCryptPasswordEncoder 加密后的密码，虽然 admin 和 sang 加密后的密码不一样，但是明文都是 123。配置完成后，使用 admin/123 或者 sang/123 就可以实现登录。本案例使用了配置在内存中的用户，一般情况下，用户信息是存储在数据库中的，因此需要在用户注册时对密码进行加密处理，代码如下：

```
1  @Service
2  public class RegService {
3      public int reg(String username, String password) {
4          BCryptPasswordEncoder encoder = new BCryptPasswordEncoder(10);
5          String encodePasswod = encoder.encode(password);
6          return saveToDb(username, encodePasswod);
7      }
8  }
```

用户将密码从前端传来之后,通过调用 BCryptPasswordEncoder 实例中的 encode 方法对密码进行加密处理,加密完成后将密文存入数据库。

10.1.9 方法安全

上文介绍的认证与授权都是基于 URL 的,开发者也可以通过注解来灵活地配置方法安全,要使用相关注解,首先要通过@EnableGlobalMethodSecurity 注解开启基于注解的安全配置:

```
1  @Configuration
2  @EnableGlobalMethodSecurity(prePostEnabled = true,securedEnabled = true)
3  public class WebSecurityConfig{
4  }
```

代码解释:

- prePostEnabled=true 会解锁@PreAuthorize 和 @PostAuthorize 两个注解,顾名思义,@PreAuthorize 注解会在方法执行前进行验证,而@PostAuthorize 注解在方法执行后进行验证。
- securedEnabled=true 会解锁@Secured 注解。

开启注解安全配置后,接下来创建一个 MethodService 进行测试,代码如下:

```
1   @Service
2   public class MethodService {
3       @Secured("ROLE_ADMIN")
4       public String admin() {
5           return "hello admin";
6       }
7       @PreAuthorize("hasRole('ADMIN') and hasRole('DBA')")
8       public String dba() {
9           return "hello dba";
10      }
11      @PreAuthorize("hasAnyRole('ADMIN','DBA','USER')")
12      public String user() {
13          return "user";
14      }
15  }
```

代码解释：
- @Secured("ROLE_ADMIN")注解表示访问该方法需要 ADMIN 角色，注意这里需要在角色前加一个前缀"ROLE_"。
- @PreAuthorize("hasRole('ADMIN') and hasRole('DBA')")注解表示访问该方法既需要 ADMIN 角色又需要 DBA 角色。
- @PreAuthorize("hasAnyRole('ADMIN','DBA','USER')")表示访问该方法需要 ADMIN、DBA 或 USER 角色。
- @PreAuthorize 和@PostAuthorize 中都可以使用基于表达式的语法。

最后，在 Controller 中注入 Service 并调用 Service 中的方法进行测试，这里比较简单，读者可以自行测试。

10.2　基于数据库的认证

上文向读者介绍的认证数据都是定义在内存中的，在真实项目中，用户的基本信息以及角色等都存储在数据库中，因此需要从数据库中获取数据进行认证。本节将向读者介绍如何使用数据库中的数据进行认证和授权。

1. 设计数据表

首先需要设计一个基本的用户角色表，如图 10-6 所示。一共三张表，分别是用户表、角色表以及用户角色关联表。为了方便测试，预置几条测试数据，如图 10-7 所示。

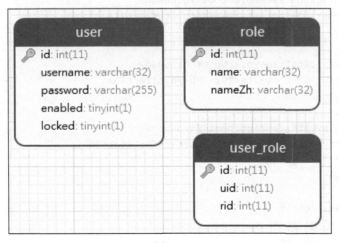

图 10-6

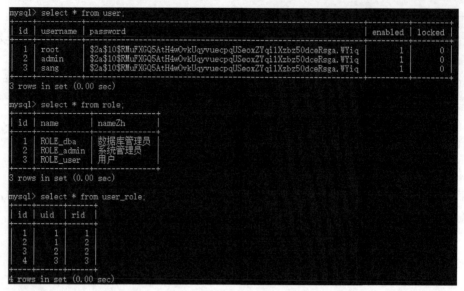

图 10-7

> **注　意**
>
> 角色名有一个默认的前缀 "ROLE_"。

2. 创建项目

MyBatis 灵活，JPA 便利，本案例选择前者，因此创建 Spring Boot Web 项目添加如下依赖：

```
1   <dependency>
2       <groupId>org.springframework.boot</groupId>
3       <artifactId>spring-boot-starter-security</artifactId>
4   </dependency>
5   <dependency>
6       <groupId>org.springframework.boot</groupId>
7       <artifactId>spring-boot-starter-web</artifactId>
8   </dependency>
9   <dependency>
10      <groupId>org.mybatis.spring.boot</groupId>
11      <artifactId>mybatis-spring-boot-starter</artifactId>
12      <version>1.3.2</version>
13  </dependency>
14  <dependency>
15      <groupId>mysql</groupId>
16      <artifactId>mysql-connector-java</artifactId>
17      <scope>runtime</scope>
18  </dependency>
19  <dependency>
20      <groupId>com.alibaba</groupId>
21      <artifactId>druid</artifactId>
22      <version>1.1.10</version>
23  </dependency>
```

3. 配置数据库

在 application.properties 中进行数据库连接配置：

```
spring.datasource.type=com.alibaba.druid.pool.DruidDataSource
spring.datasource.username=root
spring.datasource.password=root
spring.datasource.url=jdbc:mysql:///security
```

4. 创建实体类

分别创建角色表和用户表对应的实体类，代码如下：

```java
public class Role {
    private Integer id;
    private String name;
    private String nameZh;
    //省略 getter/setter
}
public class User implements UserDetails {
    private Integer id;
    private String username;
    private String password;
    private Boolean enabled;
    private Boolean locked;
    private List<Role> roles;
    @Override
    public Collection<? extends GrantedAuthority> getAuthorities() {
        List<SimpleGrantedAuthority> authorities = new ArrayList<>();
        for (Role role : roles) {
            authorities.add(new SimpleGrantedAuthority(role.getName()));
        }
        return authorities;
    }
    @Override
    public String getPassword() {
        return password;
    }
    @Override
    public String getUsername() {
        return username;
    }
    @Override
    public boolean isAccountNonExpired() {
        return true;
    }
    @Override
    public boolean isAccountNonLocked() {
        return !locked;
    }
    @Override
    public boolean isCredentialsNonExpired() {
```

```
40          return true;
41      }
42      @Override
43      public boolean isEnabled() {
44          return enabled;
45      }
46      //省略getter/setter
47  }
```

代码解释:

- 用户实体类需要实现 UserDetails 接口,并实现该接口中的 7 个方法,如表 10-1 所示。

表 10-1 UserDetails 接口的 7 个方法

方法名	解释
getAuthorities();	获取当前用户对象所具有的角色信息
getPassword();	获取当前用户对象的密码
getUsername();	获取当前用户对象的用户名
isAccountNonExpired();	当前账户是否未过期
isAccountNonLocked();	当前账户是否未锁定
isCredentialsNonExpired();	当前账户密码是否未过期
isEnabled();	当前账户是否可用

- 用户根据实际情况设置这 7 个方法的返回值。因为默认情况下不需要开发者自己进行密码角色等信息的比对,开发者只需要提供相关信息即可,例如 getPassword()方法返回的密码和用户输入的登录密码不匹配,会自动抛出 BadCredentialsException 异常,isAccountNonExpired()方法返回了 false,会自动抛出 AccountExpiredException 异常,因此对开发者而言,只需要按照数据库中的数据在这里返回相应的配置即可。本案例因为数据库中只有 enabled 和 locked 字段,故账户未过期和密码未过期两个方法都返回 true。
- getAuthorities()方法用来获取当前用户所具有的角色信息,本案例中,用户所具有的角色存储在 roles 属性中,因此该方法直接遍历 roles 属性,然后构造 SimpleGrantedAuthority 集合并返回。

5. 创建 UserService

接下来创建 UserService,代码如下:

```
1   @Service
2   public class UserService implements UserDetailsService {
3       @Autowired
4       UserMapper userMapper;
5       @Override
6       public UserDetails loadUserByUsername(String username) throws
7   UsernameNotFoundException {
8           User user = userMapper.loadUserByUsername(username);
9           if (user == null) {
10              throw new UsernameNotFoundException("账户不存在!");
```

```
11        }
12        user.setRoles(userMapper.getUserRolesByUid(user.getId()));
13        return user;
14    }
15 }
```

代码解释：

- 定义 UserService 实现 UserDetailsService 接口，并实现该接口中的 loadUserByUsername 方法，该方法的参数就是用户登录时输入的用户名，通过用户名去数据库中查找用户，如果没有查找到用户，就抛出一个账户不存在的异常，如果查找到了用户，就继续查找该用户所具有的角色信息，并将获取到的 user 对象返回，再由系统提供的 DaoAuthenticationProvider 类去比对密码是否正确。
- loadUserByUsername 方法将在用户登录时自动调用。

当然，这里还涉及 UserMapper 和 UserMapper.xml，相关源码如下：

```
1  @Mapper
2  public interface UserMapper {
3      User loadUserByUsername(String username);
4      List<Role> getUserRolesByUid(Integer id);
5  }
6  <!DOCTYPE mapper
7          PUBLIC "-//mybatis.org//DTD Mapper 3.0//EN"
8  "http://mybatis.org/dtd/mybatis-3-mapper.dtd">
9  <mapper namespace="org.sang.security02.mapper.UserMapper">
10 <select id="loadUserByUsername" resultType="org.sang.security02.model.User">
11     select * from user where username=#{username}
12 </select>
13 <select id="getUserRolesByUid" resultType="org.sang.security02.model.Role">
14     select * from role r,user_role ur where r.id=ur.rid and ur.uid=#{id}
15 </select>
16 </mapper>
```

6. 配置 Spring Security

接下来对 Spring Security 进行配置，代码如下：

```
1  @Configuration
2  public class WebSecurityConfig extends WebSecurityConfigurerAdapter {
3      @Autowired
4      UserService userService;
5      @Bean
6      PasswordEncoder passwordEncoder() {
7          return new BCryptPasswordEncoder();
8      }
9      @Override
10     protected void configure(AuthenticationManagerBuilder auth) throws Exception {
11         auth.userDetailsService(userService);
12     }
13
```

```
14      @Override
15      protected void configure(HttpSecurity http) throws Exception {
16          http.authorizeRequests()
17                  .antMatchers("/admin/**").hasRole("admin")
18                  .antMatchers("/db/**").hasRole("dba")
19                  .antMatchers("/user/**").hasRole("user")
20                  .anyRequest().authenticated()
21                  .and()
22                  .formLogin()
23                  .loginProcessingUrl("/login").permitAll()
24                  .and()
25                  .csrf().disable();
26      }
27  }
```

这里大部分配置和 10.1 节介绍的一致，唯一不同的是没有配置内存用户，而是将刚刚创建好的 UserService 配置到 AuthenticationManagerBuilder 中。

配置完成后，接下来就可以创建 Controller 进行测试了，测试方式与 10.1 节一致，这里不再赘述。

10.3 高级配置

10.3.1 角色继承

在 10.2 节的案例中定义了三种角色，但是这三种角色之间不具备任何关系，一般来说角色之间是有关系的，例如 ROLE_admin 一般既具有 admin 的权限，又具有 user 的权限。那么如何配置这种角色继承关系呢？在 Spring Security 中只需要开发者提供一个 RoleHierarchy 即可。以 10.2 中的案例为例，假设 ROLE_dba 是终极大 Boss，具有所有的权限，ROLE_admin 具有 ROLE_user 的权限，ROLE_user 则是一个公共角色，即 ROLE_admin 继承 ROLE_user、ROLE_dba 继承 ROLE_admin，要描述这种继承关系，只需要开发者在 Spring Security 的配置类中提供一个 RoleHierarchy 即可，代码如下：

```
1  @Bean
2  RoleHierarchy roleHierarchy() {
3      RoleHierarchyImpl roleHierarchy = new RoleHierarchyImpl();
4      String hierarchy = "ROLE_dba > ROLE_admin ROLE_admin > ROLE_user";
5      roleHierarchy.setHierarchy(hierarchy);
6      return roleHierarchy;
7  }
```

配置完 RoleHierarchy 之后，具有 ROLE_dba 角色的用户就可以访问所有资源了，具有 ROLE_admin 角色的用户也可以访问具有 ROLE_user 角色才能访问的资源。

10.3.2 动态配置权限

使用 HttpSecurity 配置的认证授权规则还是不够灵活，无法实现资源和角色之间的动态调整，要实现动态配置 URL 权限，就需要开发者自定义权限配置，配置步骤如下（本案例在 10.2 节案例的基础上完成）。

1. 数据库设计

这里的数据库在 10.2 节数据库的基础上再增加一张资源表和资源角色关联表，如图 10-8 所示。资源表中定义了用户能够访问的 URL 模式，资源角色表则定义了访问该模式的 URL 需要什么样的角色。

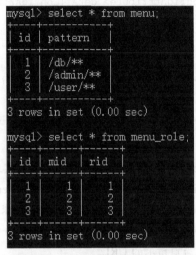

图 10-8

2. 自定义 FilterInvocationSecurityMetadataSource

要实现动态配置权限，首先要自定义 FilterInvocationSecurityMetadataSource，Spring Security 中通过 FilterInvocationSecurityMetadataSource 接口中的 getAttributes 方法来确定一个请求需要哪些角色，FilterInvocationSecurityMetadataSource 接口的默认实现类是 DefaultFilterInvocationSecurityMetadataSource，参考 DefaultFilterInvocationSecurityMetadataSource 的实现，开发者可以定义自己的 FilterInvocationSecurityMetadataSource，代码如下：

```
1   @Component
2   public class CustomFilterInvocationSecurityMetadataSource
3           implements FilterInvocationSecurityMetadataSource {
4       AntPathMatcher antPathMatcher = new AntPathMatcher();
5       @Autowired
6       MenuMapper menuMapper;
7       @Override
8       public Collection<ConfigAttribute> getAttributes(Object object)
9               throws IllegalArgumentException {
10          String requestUrl = ((FilterInvocation) object).getRequestUrl();
11          List<Menu> allMenus = menuMapper.getAllMenus();
12          for (Menu menu : allMenus) {
```

```
13                if (antPathMatcher.match(menu.getPattern(), requestUrl)) {
14                    List<Role> roles = menu.getRoles();
15                    String[] roleArr = new String[roles.size()];
16                    for (int i = 0; i < roleArr.length; i++) {
17                        roleArr[i] = roles.get(i).getName();
18                    }
19                    return SecurityConfig.createList(roleArr);
20                }
21            }
22            return SecurityConfig.createList("ROLE_LOGIN");
23        }
24        @Override
25        public Collection<ConfigAttribute> getAllConfigAttributes() {
26            return null;
27        }
28        @Override
29        public boolean supports(Class<?> clazz) {
30            return FilterInvocation.class.isAssignableFrom(clazz);
31        }
32    }
```

代码解释：

- 开发者自定义 FilterInvocationSecurityMetadataSource 主要实现该接口中的 getAttributes 方法，该方法的参数是一个 FilterInvocation，开发者可以从 FilterInvocation 中提取出当前请求的 URL，返回值是 Collection<ConfigAttribute>，表示当前请求 URL 所需的角色。
- 第 4 行创建一个 AntPathMatcher，主要用来实现 ant 风格的 URL 匹配。
- 第 10 行从参数中提取出当前请求的 URL。
- 第 11 行从数据库中获取所有的资源信息，即本案例中的 menu 表以及 menu 所对应的 role，在真实项目环境中，开发者可以将资源信息缓存在 Redis 或者其他缓存数据库中。
- 第 12~21 行遍历资源信息，遍历过程中获取当前请求的 URL 所需要的角色信息并返回。如果当前请求的 URL 在资源表中不存在相应的模式，就假设该请求登录后即可访问，即直接返回 ROLE_LOGIN。
- getAllConfigAttributes 方法用来返回所有定义好的权限资源，Spring Security 在启动时会校验相关配置是否正确，如果不需要校验，那么该方法直接返回 null 即可。
- supports 方法返回类对象是否支持校验。

3. 自定义 AccessDecisionManager

当一个请求走完 FilterInvocationSecurityMetadataSource 中的 getAttributes 方法后，接下来就会来到 AccessDecisionManager 类中进行角色信息的比对，自定义 AccessDecisionManager 如下：

```
1   @Component
2   public class CustomAccessDecisionManager implements AccessDecisionManager {
3       @Override
4       public void decide(Authentication auth,
5                          Object object,
6                          Collection<ConfigAttribute> ca){
```

```
 7              Collection<? extends GrantedAuthority> auths = auth.getAuthorities();
 8              for (ConfigAttribute configAttribute : ca) {
 9                  if ("ROLE_LOGIN".equals(configAttribute.getAttribute())
10  && auth instanceof UsernamePasswordAuthenticationToken) {
11                      return;
12                  }
13                  for (GrantedAuthority authority : auths) {
14                      if (configAttribute.getAttribute().equals(authority.getAuthority()))
15  {
16                          return;
17                      }
18                  }
19              }
20              throw new AccessDeniedException("权限不足");
21          }
22
23          @Override
24          public boolean supports(ConfigAttribute attribute) {
25              return true;
26          }
27
28          @Override
29          public boolean supports(Class<?> clazz) {
30              return true;
31          }
    }
```

代码解释:

- 自定义 AccessDecisionManager 并重写 decide 方法, 在该方法中判断当前登录的用户是否具备当前请求 URL 所需要的角色信息, 如果不具备, 就抛出 AccessDeniedException 异常, 否则不做任何事即可。
- decide 方法有三个参数, 第一个参数包含当前登录用户的信息; 第二个参数则是一个 FilterInvocation 对象, 可以获取当前请求对象等; 第三个参数就是 FilterInvocationSecurityMetadataSource 中的 getAttributes 方法的返回值, 即当前请求 URL 所需要的角色。
- 第 7~19 行进行角色信息对比, 如果需要的角色是 ROLE_LOGIN, 说明当前请求的 URL 用户登录后即可访问, 如果 auth 是 UsernamePasswordAuthenticationToken 的实例, 那么说明当前用户已登录, 该方法到此结束, 否则进入正常的判断流程, 如果当前用户具备当前请求需要的角色, 那么方法结束。

当然, 本案例还涉及 MenuMapper 和 MenuMapper.xml, 实现如下:

```
1  @Mapper
2  public interface MenuMapper {
3      List<Menu> getAllMenus();
4  }
5  <!DOCTYPE mapper
```

```xml
 6            PUBLIC "-//mybatis.org//DTD Mapper 3.0//EN"
 7   "http://mybatis.org/dtd/mybatis-3-mapper.dtd">
 8   <mapper namespace="org.sang.security02.mapper.MenuMapper">
 9   <resultMap id="BaseResultMap" type="org.sang.security02.model.Menu">
10   <id property="id" column="id"/>
11   <result property="pattern" column="pattern"/>
12   <collection property="roles" ofType="org.sang.security02.model.Role">
13   <id property="id" column="rid"/>
14   <result property="name" column="rname"/>
15   <result property="nameZh" column="rnameZh"/>
16   </collection>
17   </resultMap>
18   <select id="getAllMenus" resultMap="BaseResultMap">
19       SELECT m.*,r.id AS rid,r.name AS rname,r.nameZh AS rnameZh FROM menu m LEFT
20   JOIN menu_role mr ON m.`id`=mr.`mid` LEFT JOIN role r ON mr.`rid`=r.`id`
21   </select>
22   </mapper>
```

4. 配置

最后，在 Spring Security 中配置如上两个自定义类，部分源码如下：

```java
 1  @Configuration
 2  public class WebSecurityConfig extends WebSecurityConfigurerAdapter {
 3      @Override
 4      protected void configure(HttpSecurity http) throws Exception {
 5          http.authorizeRequests()
 6                  .withObjectPostProcessor(new
 7  ObjectPostProcessor<FilterSecurityInterceptor>() {
 8                      @Override
 9                      public <O extends FilterSecurityInterceptor> O postProcess(O object) {
10                          object.setSecurityMetadataSource(cfisms());
11                          object.setAccessDecisionManager(cadm());
12                          return object;
13                      }
14                  })
15                  .and()
16                  .formLogin()
17                  .loginProcessingUrl("/login").permitAll()
18                  .and()
19                  .csrf().disable();
20      }
21      @Bean
22      CustomFilterInvocationSecurityMetadataSource cfisms() {
23          return new CustomFilterInvocationSecurityMetadataSource();
24      }
25      @Bean
26      CustomAccessDecisionManager cadm() {
27          return new CustomAccessDecisionManager();
28      }
    }
```

代码解释:

- 本案例 WebSecurityConfig 类的定义是对 10.2 节中 WebSecurityConfig 定义的补充，主要是修改了 configure(HttpSecurity http)方法的实现并添加了两个 Bean。
- 第 9、10 行，在定义 FilterSecurityInterceptor 时，将我们自定义的两个实例设置进去即可。

经过上面的配置，我们已经实现了动态配置权限，权限和资源的关系可以在 menu_role 表中动态调整。测试案例可以参考 10.1 节的测试案例，这里不再赘述。

10.4　OAuth 2

10.4.1　OAuth 2 简介

OAuth 是一个开放标准，该标准允许用户让第三方应用访问该用户在某一网站上存储的私密资源（如头像、照片、视频等），而在这个过程中无须将用户名和密码提供给第三方应用。实现这一功能是通过提供一个令牌（token），而不是用户名和密码来访问他们存放在特定服务提供者的数据。每一个令牌授权一个特定的网站在特定的时段内访问特定的资源。这样，OAuth 让用户可以授权第三方网站灵活地访问存储在另外一些资源服务器的特定信息，而非所有内容。例如，用户想通过 QQ 登录知乎，这时知乎就是一个第三方应用，知乎要访问用户的一些基本信息就需要得到用户的授权，如果用户把自己的 QQ 用户名和密码告诉知乎，那么知乎就能访问用户的所有数据，并且只有用户修改密码才能收回授权，这种授权方式安全隐患很大，如果使用 OAuth，就能很好地解决这一问题。

采用令牌的方式可以让用户灵活地对第三方应用授权或者收回权限。OAuth 2 是 OAuth 协议的下一版本，但不向下兼容 OAuth 1.0。OAuth 2 关注客户端开发者的简易性，同时为 Web 应用、桌面应用、移动设备、起居室设备提供专门的认证流程。传统的 Web 开发登录认证一般都是基于 Session 的，但是在前后端分离的架构中继续使用 Session 会有许多不便，因为移动端（Android、iOS、微信小程序等）要么不支持 Cookie（微信小程序），要么使用非常不便，对于这些问题，使用 OAuth 2 认证都能解决。

10.4.2　OAuth 2 角色

要了解 OAuth 2，需要先了解 OAuth 2 中几个基本的角色。

- 资源所有者：资源所有者即用户，具有头像、照片、视频等资源。
- 客户端：客户端即第三方应用，例如上文提到的知乎。
- 授权服务器：授权服务器用来验证用户提供的信息是否正确，并返回一个令牌给第三方应用。
- 资源服务器：资源服务器是提供给用户资源的服务器，例如头像、照片、视频等。

一般来说，授权服务器和资源服务器可以是同一台服务器。

10.4.3　OAuth 2 授权流程

OAuth 2 的授权流程到底是什么样的呢？如图 10-9 所示。

这是 OAuth 2 一个大致的授权流程图，具体步骤如下：

步骤01 客户端（第三方应用）向用户请求授权。

步骤02 用户单击客户端所呈现的服务授权页面上的同意授权按钮后，服务端返回一个授权许可凭证给客户端。

步骤03 客户端拿着授权许可凭证去授权服务器申请令牌。

步骤04 授权服务器验证信息无误后，发放令牌给客户端。

步骤05 客户端拿着令牌去资源服务器访问资源。

步骤06 资源服务器验证令牌无误后开放资源。

这是一个大致的流程，因为 OAuth 2 中有 4 种不同的授权模式，每种授权模式的授权流程又会有差异，基本流程如图 10-9 所示。

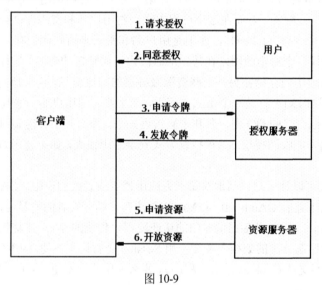

图 10-9

10.4.4　授权模式

OAuth 协议的授权模式共分为 4 种，分别说明如下。

- 授权码模式：授权码模式（authorization code）是功能最完整、流程最严谨的授权模式。它的特点就是通过客户端的服务器与授权服务器进行交互，国内常见的第三方平台登录功能基本都是使用这种模式。
- 简化模式：简化模式不需要客户端服务器参与，直接在浏览器中向授权服务器申请令牌，一般若网站是纯静态页面，则可以采用这种方式。
- 密码模式：密码模式是用户把用户名密码直接告诉客户端，客户端使用这些信息向授权服务器申请令牌。这需要用户对客户端高度信任，例如客户端应用和服务提供商是同一家公司。

- 客户端模式：客户端模式是指客户端使用自己的名义而不是用户的名义向服务提供者申请授权。严格来说，客户端模式并不能算作 OAuth 协议要解决的问题的一种解决方案，但是，对于开发者而言，在一些前后端分离应用或者为移动端提供的认证授权服务器上使用这种模式还是非常方便的。

这 4 种模式各有千秋，分别适用于不同的开发场景，开发者要根据实际情况进行选择。

10.4.5 实践

本案例要介绍的是在前后端分离应用（或者为移动端、微信小程序等）提供的认证服务器中如何搭建 OAuth 服务，因此主要介绍密码模式。搭建步骤如下。

1. 创建项目，添加依赖

创建 Spring Boot Web 项目，添加如下依赖：

```
 1  <dependency>
 2      <groupId>org.springframework.boot</groupId>
 3      <artifactId>spring-boot-starter-security</artifactId>
 4  </dependency>
 5  <dependency>
 6      <groupId>org.springframework.boot</groupId>
 7      <artifactId>spring-boot-starter-data-redis</artifactId>
 8      <exclusions>
 9          <exclusion>
10              <groupId>io.lettuce</groupId>
11              <artifactId>lettuce-core</artifactId>
12          </exclusion>
13      </exclusions>
14  </dependency>
15  <dependency>
16      <groupId>redis.clients</groupId>
17      <artifactId>jedis</artifactId>
18  </dependency>
19  <dependency>
20      <groupId>org.springframework.boot</groupId>
21      <artifactId>spring-boot-starter-web</artifactId>
22  </dependency>
23  <dependency>
24      <groupId>org.springframework.security.oauth</groupId>
25      <artifactId>spring-security-oauth2</artifactId>
26      <version>2.3.3.RELEASE</version>
27  </dependency>
```

由于 Spring Boot 中的 OAuth 协议是在 Spring Security 的基础上完成的，因此首先要添加 Spring Security 依赖，要用到 OAuth 2，因此添加 OAuth 2 相关依赖，令牌可以存储在 Redis 缓存服务器上，同时 Redis 具有过期等功能，很适合令牌的存储，因此也加入 Redis 依赖。

项目创建成功后，接下来在 application.properties 中配置一下 Redis 服务器的连接信息，代码如下：

```
1  spring.redis.database=0
2  spring.redis.host=192.168.66.129
3  spring.redis.port=6379
4  spring.redis.password=123@456
5  spring.redis.jedis.pool.max-active=8
6  spring.redis.jedis.pool.max-idle=8
7  spring.redis.jedis.pool.max-wait=-1ms
8  spring.redis.jedis.pool.min-idle=0
```

Redis 配置可以参考 6.1 节，这里不再赘述。

2. 配置授权服务器

授权服务器和资源服务器可以是同一台服务器，也可以是不同服务器，本案例中假设是同一台服务器，通过不同的配置分别开启授权服务器和资源服务器，首先是授权服务器：

```
1   @Configuration
2   @EnableAuthorizationServer
3   public class AuthorizationServerConfig
4           extends AuthorizationServerConfigurerAdapter {
5       @Autowired
6       AuthenticationManager authenticationManager;
7       @Autowired
8       RedisConnectionFactory redisConnectionFactory;
9       @Autowired
10      UserDetailsService userDetailsService;
11      @Bean
12      PasswordEncoder passwordEncoder() {
13          return new BCryptPasswordEncoder();
14      }
15      @Override
16      public void configure(ClientDetailsServiceConfigurer clients)
17              throws Exception {
18          clients.inMemory()
19          .withClient("password")
20          .authorizedGrantTypes("password", "refresh_token")
21          .accessTokenValiditySeconds(1800)
22          .resourceIds("rid")
23          .scopes("all")
24          .secret("$2a$10$RMuFXGQ5AtH4wOvkUqyvuecpqUSeoxZYqilXzbz50dceRsga.WYiq");
25      }
26      @Override
27      public void configure(AuthorizationServerEndpointsConfigurer endpoints)
28              throws Exception {
29          endpoints.tokenStore(new RedisTokenStore(redisConnectionFactory))
30                  .authenticationManager(authenticationManager)
31                  .userDetailsService(userDetailsService);
32      }
33      @Override
```

第 10 章　Spring Boot 安全管理

```
34        public void configure(AuthorizationServerSecurityConfigurer security)
35                throws Exception {
36            security.allowFormAuthenticationForClients();
37        }
38    }
```

代码解释：

- 自定义类继承自 AuthorizationServerConfigurerAdapter，完成对授权服务器的配置，然后通过 @EnableAuthorizationServer 注解开启授权服务器。
- 第 5、6 行注入了 AuthenticationManager，该对象将用来支持 password 模式。
- 第 7、8 行注入了 RedisConnectionFactory，该对象将用来完成 Redis 缓存，将令牌信息存储到 Redis 缓存中。
- 第 9、10 行注入了 UserDetailsService，该对象将为刷新 token 提供支持。
- 第 11~14 行提供一个 PasswordEncoder，这个和前文中的配置一样，不再赘述。
- 第 19~24 行配置 password 授权模式，authorizedGrantTypes 表示 OAuth 2 中的授权模式为 "password" 和 "refresh_token" 两种，在标准的 OAuth 2 协议中，授权模式并不包括 "refresh_token"，但是在 Spring Security 的实现中将其归为一种，因此如果要实现 access_token 的刷新，就需要添加这样一种授权模式；accessTokenValiditySeconds 方法配置了 access_token 的过期时间；resourceIds 配置了资源 id；secret 方法配置了加密后的密码，明文是 123。
- 第 29 行配置了令牌的存储，AuthenticationManager 和 UserDetailsService 主要用于支持 password 模式以及令牌的刷新。
- 第 36 行的配置表示支持 client_id 和 client_secret 做登录认证。

3. 配置资源服务器

接下来配置资源服务器，代码如下：

```
1   @Configuration
2   @EnableResourceServer
3   public class ResourceServerConfig extends ResourceServerConfigurerAdapter {
4       @Override
5       public void configure(ResourceServerSecurityConfigurer resources)
6               throws Exception {
7           resources.resourceId("rid").stateless(true);
8       }
9       @Override
10      public void configure(HttpSecurity http) throws Exception {
11          http.authorizeRequests()
12                  .antMatchers("/admin/**").hasRole("admin")
13                  .antMatchers("/user/**").hasRole("user")
14                  .anyRequest().authenticated();
15      }
16  }
```

代码解释：

- 自定义类继承自 ResourceServerConfigurerAdapter，并添加 @EnableResourceServer 注解开启资

源服务器配置。
- 第 7 行配置资源 id，这里的资源 id 和授权服务器中的资源 id 一致，然后设置这些资源仅基于令牌认证。
- 第 11~14 行配置 HttpSecurity，这和 10.1 节介绍的配置一致，不再赘述。

4. 配置 Security

接下来配置 Spring Security，代码如下：

```
1   @Configuration
2   public class WebSecurityConfig extends WebSecurityConfigurerAdapter {
3       @Bean
4       @Override
5       public AuthenticationManager authenticationManagerBean() throws Exception {
6           return super.authenticationManagerBean();
7       }
8       @Bean
9       @Override
10      protected UserDetailsService userDetailsService() {
11          return super.userDetailsService();
12      }
13      @Override
14      protected void configure(AuthenticationManagerBuilder auth) throws Exception {
15          auth.inMemoryAuthentication()
16              .withUser("admin")
17              .password("$2a$10$RMuFXGQ5AtH4wOvkUqyvuecpqUSeoxZYqilXzbz50dceRsga.WYiq")
18              .roles("admin")
19              .and()
20              .withUser("sang")
21              .password("$2a$10$RMuFXGQ5AtH4wOvkUqyvuecpqUSeoxZYqilXzbz50dceRsga.WYiq")
22              .roles("user");
23      }
24      @Override
25      protected void configure(HttpSecurity http) throws Exception {
26          http.antMatcher("/oauth/**").authorizeRequests()
27                  .antMatchers("/oauth/**").permitAll()
28                  .and().csrf().disable();
29      }
30  }
```

这里 Spring Security 的配置基本上和前文一致，唯一不同的是多了两个 Bean，这里两个 Bean 将注入授权服务器配置类中使用。另外，这里的 HttpSecurity 配置主要是配置 "/oauth/**" 模式的 URL，这一类的请求直接放行。在 Spring Security 配置和资源服务器配置中，一共涉及两个 HttpSecurity，其中 Spring Security 中的配置优先级高于资源服务器中的配置，即请求地址先经过 Spring Security 的 HttpSecurity，再经过资源服务器的 HttpSecurity。

5. 测试验证

首先创建三个简单的请求地址，代码如下：

```
1   @RestController
2   public class HelloController {
3       @GetMapping("/admin/hello")
4       public String admin() {
5           return "Hello admin!";
6       }
7       @GetMapping("/user/hello")
8       public String user() {
9           return "Hello user!";
10      }
11      @GetMapping("/hello")
12      public String hello() {
13          return "hello";
14      }
15  }
```

根据前文的配置，要请求这三个地址，分别需要 admin 角色、user 角色以及登录后访问。

所有都配置完成后，启动 Redis 服务器，再启动 Spring Boot 项目，首先发送一个 POST 请求获取 token，请求地址如下（注意这里是一个 POST 请求，为了显示方便，将参数写在地址栏中）：

http://localhost:8080/oauth/token?username=sang&password=123&grant_type=password&client_id=password&scope=all&client_secret=123

请求地址中包含的参数有用户名、密码、授权模式、客户端 id、scope 以及客户端密码，基本就是授权服务器中所配置的数据，请求结果如图 10-10 所示。

```
{
    "access_token": "918f7927-6144-4b47-ac5e-df7ee6a03fb6",
    "token_type": "bearer",
    "refresh_token": "330335cd-a1a1-44b7-8d9d-1c531dd0e82e",
    "expires_in": 1799,
    "scope": "all"
}
```

图 10-10

返回结果有 access_token、token_type、refresh_token、expires_in 以及 scope，其中 access_token 是获取其他资源时要用的令牌，refresh_token 用来刷新令牌，expires_in 表示 access_token 的过期时间，当 access_token 过期后，使用 refresh_token 重新获取新的 access_token（前提是 refresh_token 未过期），请求地址如下（注意这里也是 POST 请求）：

http://localhost:8080/oauth/token?grant_type=refresh_token&refresh_token=6b80de74-e264-4ca5-b9cc-20e8a6fa486d&client_id=password&client_secret=123

获取新的 access_token 时需要携带上 refresh_token，同时授权模式设置为 refresh_token，在获取的结果中 access_token 会变化，同时 access_token 有效期也会变化，如图 10-11 所示。

```
{
    "access_token": "136ea275-ae18-4b36-8bfd-0b4963bb11af",
    "token_type": "bearer",
    "refresh_token": "330335cd-a1a1-44b7-8d9d-1c531dd0e82e",
    "expires_in": 1799,
    "scope": "all"
}
```

图 10-11

接下来访问所有资源，携带上 access_token 参数即可，例如 "/user/hello" 接口：
http://localhost:8080/user/hello?access_token=136ea275-ae18-4b36-8bfd-0b4963bb11af
访问结果如图 10-12 所示。

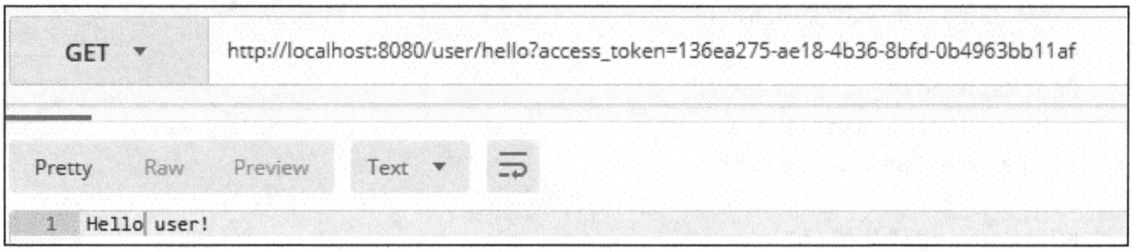

图 10-12

如果非法访问一个资源，例如 sang 用户访问 "/admin/hello" 接口，结果如图 10-13 所示。

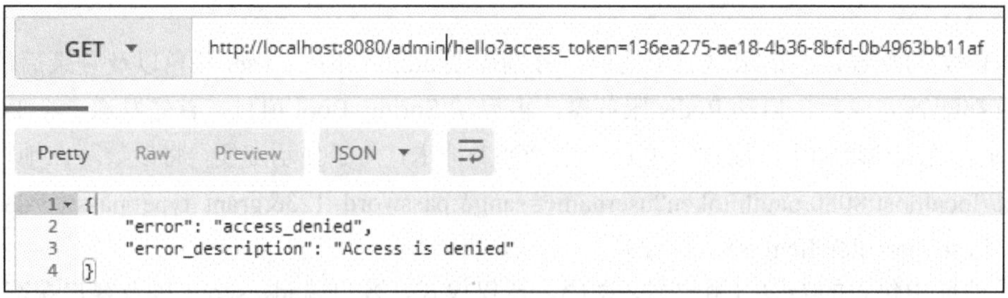

图 10-13

最后，再来看一下 Redis 中的数据，如图 10-14 所示。

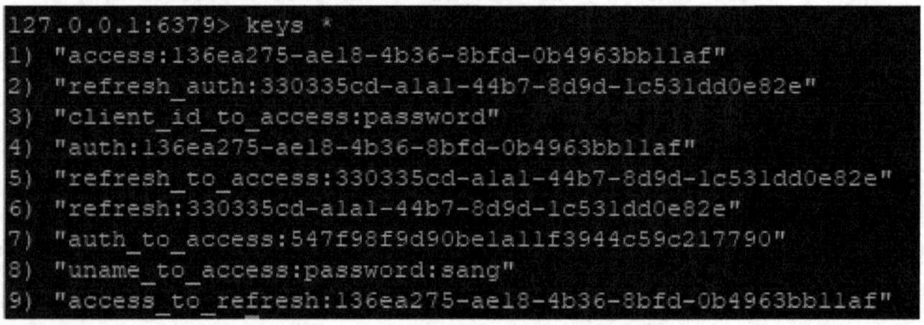

图 10-14

到此，一个 password 模式的 OAuth 认证体系就搭建成功了。

OAuth 中的认证模式有 4 种，读者需要结合自己开发的实际情况选择其中一种，本案例介绍的是在前后端分离应用中常用的 password 模式，其他的授权模式也都有自己的使用场景。

整体来说，Spring Security OAuth 2 的使用还是较复杂的，配置也比较烦琐，如果开发者的应用场景比较简单，完全可以按照前文介绍的授权流程自己搭建 OAuth 2 认证体系。

10.5 Spring Boot 整合 Shiro

10.5.1 Shiro 简介

Apache Shiro 是一个开源的轻量级的 Java 安全框架,它提供身份验证、授权、密码管理以及会话管理等功能。相对于 Spring Security,Shiro 框架更加直观、易用,同时也能提供健壮的安全性。在传统的 SSM 框架中,手动整合 Shiro 的配置步骤还是比较多的,针对 Spring Boot,Shiro 官方提供了 shiro-spring-boot-web-starter 用来简化 Shiro 在 Spring Boot 中的配置。下面向读者介绍 shiro-spring-boot-web-starter 的使用步骤。

10.5.2 整合 Shiro

1. 创建项目

首先创建一个普通的 Spring Boot Web 项目,添加 Shiro 依赖以及页面模板依赖,代码如下:

```
1  <dependency>
2    <groupId>org.apache.shiro</groupId>
3    <artifactId>shiro-spring-boot-web-starter</artifactId>
4    <version>1.4.0</version>
5  </dependency>
6  <dependency>
7    <groupId>org.springframework.boot</groupId>
8    <artifactId>spring-boot-starter-thymeleaf</artifactId>
9  </dependency>
10 <dependency>
11   <groupId>com.github.theborakompanioni</groupId>
12   <artifactId>thymeleaf-extras-shiro</artifactId>
13   <version>2.0.0</version>
14 </dependency>
```

注意这里不需要添加 spring-boot-starter-web 依赖,shiro-spring-boot-web-starter 中已经依赖了 spring-boot-starter-web。同时,本案例使用 Thymeleaf 模板,因此添加 Thymeleaf 依赖,另外,为了在 Thymeleaf 中使用 shiro 标签,因此引入了 thymeleaf-extras-shiro 依赖。

2. Shiro 基本配置

首先在 application.properties 中配置 Shiro 的基本信息,代码如下:

```
1  shiro.enabled=true
2  shiro.web.enabled=true
3  shiro.loginUrl=/login
4  shiro.successUrl=/index
5  shiro.unauthorizedUrl=/unauthorized
6  shiro.sessionManager.sessionIdUrlRewritingEnabled=true
7  shiro.sessionManager.sessionIdCookieEnabled=true
```

代码解释：

- 第 1 行配置表示开启 Shiro 配置，默认为 true。
- 第 2 行配置表示开启 Shiro Web 配置，默认为 true。
- 第 3 行配置表示登录地址，默认为 "/loing.jsp"。
- 第 4 行配置表示登录成功地址，默认为 "/"。
- 第 5 行配置表示未获授权默认跳转地址。
- 第 6 行配置表示是否允许通过 URL 参数实现会话跟踪，如果网站支持 Cookie，可以关闭此选项，默认为 true。
- 第 7 行配置表示是否允许通过 Cookie 实现会话跟踪，默认为 true。

基本信息配置完成后，接下来在 Java 代码中配置 Shiro，提供两个最基本的 Bean 即可，代码如下：

```
@Configuration
public class ShiroConfig {
    @Bean
    public Realm realm() {
        TextConfigurationRealm realm = new TextConfigurationRealm();
        realm.setUserDefinitions("sang=123,user\n admin=123,admin");
        realm.setRoleDefinitions("admin=read,write\n user=read");
        return realm;
    }
    @Bean
    public ShiroFilterChainDefinition shiroFilterChainDefinition() {
        DefaultShiroFilterChainDefinition chainDefinition =
                new DefaultShiroFilterChainDefinition();
        chainDefinition.addPathDefinition("/login", "anon");
        chainDefinition.addPathDefinition("/doLogin", "anon");
        chainDefinition.addPathDefinition("/logout", "logout");
        chainDefinition.addPathDefinition("/**", "authc");
        return chainDefinition;
    }
    @Bean
    public ShiroDialect shiroDialect() {
        return new ShiroDialect();
    }
}
```

代码解释：

- 这里提供两个关键 Bean，一个是 Realm，另一个是 ShiroFilterChainDefinition。至于 ShiroDialect，则是为了支持在 Thymeleaf 中使用 Shiro 标签，如果不在 Thymeleaf 中使用 Shiro 标签，那么可以不提供 ShiroDialect。
- Realm 可以是自定义 Realm，也可以是 Shiro 提供的 Realm，简单起见，本案例没有配置数据库连接，这里直接配置了两个用户：sang/123 和 admin/123，分别对应角色 user 和 admin，user 具有 read 权限，admin 则具有 read、write 权限。

- ShiroFilterChainDefinition Bean 中配置了基本的过滤规则，"/login"和"/doLogin"可以匿名访问，"/logout"是一个注销登录请求，其余请求则都需要认证后才能访问。

接下来配置登录接口以及页面访问接口，代码如下：

```
@Controller
public class UserController {
    @PostMapping("/doLogin")
    public String doLogin(String username, String password, Model model) {
        UsernamePasswordToken token =
                new UsernamePasswordToken(username, password);
        Subject subject = SecurityUtils.getSubject();
        try {
            subject.login(token);
        } catch (AuthenticationException e) {
            model.addAttribute("error", "用户名或密码输入错误!");
            return "login";
        }
        return "redirect:/index";
    }
    @RequiresRoles("admin")
    @GetMapping("/admin")
    public String admin() {
        return "admin";
    }
    @RequiresRoles(value = {"admin","user"},logical = Logical.OR)
    @GetMapping("/user")
    public String user() {
        return "user";
    }
}
```

代码解释：

- 在 doLogin 方法中，首先构造一个 UsernamePasswordToken 实例，然后获取一个 Subject 对象并调用该对象中的 login 方法执行登录操作，在登录操作执行过程中，当有异常抛出时，说明登录失败，携带错误信息返回登录视图；当登录成功时，则重定向到 "/index"。
- 接下来暴露两个接口 "/admin" 和 "/user"，对于 "/admin" 接口，需要具有 admin 角色才可以访问；对于 "/user" 接口，具备 admin 角色和 user 角色其中任意一个即可访问。

对于其他不需要角色就能访问的接口，直接在 WebMvc 中配置即可，代码如下：

```
@Configuration
public class WebMvcConfig implements WebMvcConfigurer{
    @Override
    public void addViewControllers(ViewControllerRegistry registry) {
        registry.addViewController("/login").setViewName("login");
        registry.addViewController("/index").setViewName("index");
        registry.addViewController("/unauthorized").setViewName("unauthorized");
    }
```

```
9  }
```

接下来创建全局异常处理器进行全局异常处理，本案例主要是处理授权异常，代码如下：

```
1  @ControllerAdvice
2  public class ExceptionController {
3      @ExceptionHandler(AuthorizationException.class)
4      public ModelAndView error(AuthorizationException e) {
5          ModelAndView mv = new ModelAndView("unauthorized");
6          mv.addObject("error", e.getMessage());
7          return mv;
8      }
9  }
```

当用户访问未授权的资源时，跳转到 unauthorized 视图中，并携带出错信息。

配置完成后，最后在 resources/templates 目录下创建 5 个 HTML 页面进行测试。

（1）index.html，代码如下：

```
1   <!DOCTYPE html>
2   <html lang="en" xmlns:shiro="http://www.pollix.at/thymeleaf/shiro">
3   <head>
4   <meta charset="UTF-8">
5   <title>Title</title>
6   </head>
7   <body>
8   <h3>Hello, <shiro:principal/></h3>
9   <h3><a href="/logout">注销登录</a></h3>
10  <h3><a shiro:hasRole="admin" href="/admin">管理员页面</a></h3>
11  <h3><a shiro:hasAnyRoles="admin,user" href="/user">普通用户页面</a></h3>
12  </body>
13  </html>
```

index.html 是登录成功后的首页，首先展示当前登录用户的用户名，然后展示一个"注销登录"链接，若当前登录用户具备"admin"角色，则展示一个"管理员页面"的超链接；若用户具备"admin"或者"user"角色，则展示一个"普通用户页面"的超链接。注意这里导入的名称空间是 xmlns:shiro=http://www.pollix.at/thymeleaf/shiro，和 JSP 中导入的 Shiro 名称空间不一致。

（2）login.html，代码如下：

```
1   <!DOCTYPE html>
2   <html lang="en" xmlns:th="http://www.thymeleaf.org">
3   <head>
4   <meta charset="UTF-8">
5   <title>Title</title>
6   </head>
7   <body>
8   <div>
9   <form action="/doLogin" method="post">
10  <input type="text" name="username"><br>
11  <input type="password" name="password"><br>
12  <div th:text="${error}"></div>
```

```
13  <input type="submit" value="登录">
14  </form>
15  </div>
16  </body>
17  </html>
```

login.html 是一个普通的登录页面，在登录失败时通过一个 div 显示登录失败信息。

（3）user.html，代码如下：

```
1   <!DOCTYPE html>
2   <html lang="en">
3   <head>
4   <meta charset="UTF-8">
5   <title>Title</title>
6   </head>
7   <body>
8   <h1>普通用户页面</h1>
9   </body>
10  </html>
```

user.html 是一个普通的用户信息展示页面。

（4）admin.html，代码如下：

```
1   <!DOCTYPE html>
2   <html lang="en">
3   <head>
4   <meta charset="UTF-8">
5   <title>Title</title>
6   </head>
7   <body>
8   <h1>管理员页面</h1>
9   </body>
10  </html>
```

admin.html 是一个普通的管理员信息展示页面。

（5）unauthorized.html，代码如下：

```
1   <!DOCTYPE html>
2   <html lang="en" xmlns:th="http://www.thymeleaf.org">
3   <head>
4   <meta charset="UTF-8">
5   <title>Title</title>
6   </head>
7   <body>
8   <div>
9   <h3>未获授权，非法访问</h3>
10  <h3 th:text="${error}"></h3>
11  </div>
12  </body>
13  </html>
```

unauthorized.html 是一个授权失败的展示页面，该页面还会展示授权出错的信息。

3. 测试

配置完成后，启动 Spring Boot 项目，访问登录页面，分别使用 sang/123 和 admin/123 登录，结果如图 10-15、图 10-16 所示。注意，因为 sang 用户不具备 admin 角色，因此登录成功后的页面上没有前往管理员页面的超链接。

图 10-15

图 10-16

登录成功后，无论是 sang 还是 admin 用户，单击"注销登录"都会注销成功，然后回到登录页面，sang 用户因为不具备 admin 角色，因此没有"管理员页面"的超链接，无法进入管理员页面中，此时，若用户使用 sang 用户登录，然后手动在地址栏输入 http://localhost:8080/admin，则会跳转到未授权页面，如图 10-17 所示。

图 10-17

以上通过一个简单的案例向读者展示了如何在 Spring Boot 中整合 Shiro 以及如何在 Thymeleaf 中使用 Shiro 标签，一旦整合成功，接下来 Shiro 的用法就和原来的一模一样。本小节主要介绍 Spring Boot 整合 Shiro，对于 Shiro 的其他用法，读者可以参考 Shiro 官方文档，这里不再赘述。

10.6　小　　结

本章主要向读者介绍了 Spring Security 以及 Shiro 在 Spring Boot 中的使用。对于 Spring Security，有基于传统认证方式的 Session 认证，也有使用 OAuth 协议的认证。一般来说，在传统的 Web 架构中，使用 Session 认证方便快速，但是，若结合微服务、前后端分离等架构，则使用 OAuth 认证更加方便，具体使用哪一种，需要开发者根据实际情况进行取舍。而对于 Shiro，虽然功能不及 Spring Security 强大，但是简单易用，而且也能胜任大部分的中小型项目。当然，在 Spring Boot 项目中，Spring Security 的整合显然要更加容易，因此可以首选 Spring Security。如果开发团队对 Spring Security 不熟悉却熟悉 Shiro 的使用，当然也可以使用 Shiro，这个要结合具体情况来定。

第 11 章

Spring Boot 整合 WebSocket

本章概要

- 为什么需要 WebSocket
- WebSocket 简介
- Spring Boot 整合 WebSocket

11.1 为什么需要 WebSocket

在 HTTP 协议中，所有的请求都是由客户端发起的，由服务端进行响应，服务端无法向客户端推送消息，但是在一些需要即时通信的应用中，又不可避免地需要服务端向客户端推送消息，传统的解决方案主要有如下几种。

1. 轮询

轮询是最简单的一种解决方案，所谓轮询，就是客户端在固定的时间间隔下不停地向服务端发送请求，查看服务端是否有最新的数据，若服务端有最新的数据，则返回给客户端，若服务端没有，则返回一个空的 JSON 或者 XML 文档。轮询对开发人员而言实现方便，但是弊端也很明显：客户端每次都要新建 HTTP 请求，服务端要处理大量的无效请求，在高并发场景下会严重拖慢服务端的运行效率，同时服务端的资源被极大的浪费了，因此这种方式并不可取。

2. 长轮询

长轮询是传统轮询的升级版，当聪明的工程师看到轮询所存在的问题后，就开始解决问题，于是有了长轮询。不同于传统轮询，在长轮询中，服务端不是每次都会立即响应客户端的请求，只有在服务端有最新数据的时候才会立即响应客户端的请求，否则服务端会持有这个请求而不返回，直到有新数据时才返回。这种方式可以在一定程度上节省网络资源和服务器资源，但是也存在一些问题，例如：

- 如果浏览器在服务器响应之前有新数据要发送，就只能创建一个新的并发请求，或者先尝试断掉当前请求，再创建新的请求。
- TCP 和 HTTP 规范中都有连接超时一说，所以所谓的长轮询并不能一直持续，服务端和客户端的连接需要定期的连接和关闭再连接，这又增大了程序员的工作量，当然也有一些技术能够延长每次连接的时间，但毕竟是非主流解决方案。

3. Applet 和 Flash

Applet 和 Flash 都已经是明日黄花，不过在这两个技术存在的岁月里，除了可以让我们的 HTML 页面更加绚丽之外，还可以解决消息推送问题。开发者可以使用 Applet 和 Flash 来模拟全双工通信，通过创建一个只有 1 个像素点大小的透明的 Applet 或者 Flash，然后将之内嵌在网页中，再从 Applet 或者 Flash 的代码中创建一个 Socket 连接进行双向通信。这种连接方式消除了 HTTP 协议中的诸多限制，当服务器有消息发送到客户端的时候，开发者可以在 Applet 或者 Flash 中调用 JavaScript 函数将数据显示在页面上，当浏览器有数据要发送给服务器时也一样，通过 Applet 或者 Flash 来传递。这种方式真正地实现了全双工通信，不过也有问题，说明如下：

- 浏览器必须能够运行 Java 或者 Flash。
- 无论是 Applet 还是 Flash 都存在安全问题。
- 随着 HTML 5 标准被各浏览器厂商广泛支持，Flash 下架已经被提上日程（Adobe 宣布 2020 年正式停止支持 Flash）。

其实，传统的解决方案不止这三种，但是无论哪种解决方案都有自身的缺陷，于是有了 WebSocket。

11.2 WebSocket 简介

WebSocket 是一种在单个 TCP 连接上进行全双工通信的协议，已被 W3C 定为标准。使用 WebSocket 可以使得客户端和服务器之间的数据交换变得更加简单，它允许服务端主动向客户端推送数据。在 WebSocket 协议中，浏览器和服务器只需要完成一次握手，两者之间就可以直接创建持久性的连接，并进行双向数据传输。

WebSocket 使用了 HTTP/1.1 的协议升级特性，一个 WebSocket 请求首先使用非正常的 HTTP 请求以特定的模式访问一个 URL，这个 URL 有两种模式，分别是 ws 和 wss，对应 HTTP 协议中的 HTTP 和 HTTPS，在请求头中有一个 Connection:Upgrade 字段，表示客户端想要对协议进行升级，另外还有一个 Upgrade:websocket 字段，表示客户端想要将请求协议升级为 WebSocket 协议。这两个字段共同告诉服务器要将连接升级为 WebSocket 这样一种全双工协议，如果服务端同意协议升级，那么在握手完成之后，文本消息或者其他二进制消息就可以同时在两个方向上进行发送，而不需要关闭和重建连接。此时的客户端和服务端关系是对等的，它们可以互相向对方主动发送消息。和传统的解决方案相比，WebSocket 主要有如下特点：

- WebSocket 使用时需要先创建连接，这使得 WebSocket 成为一种有状态的协议，在之后的通信过程中可以省略部分状态信息（例如身份认证等）。
- WebSocket 连接在端口 80（ws）或者 443（wss）上创建，与 HTTP 使用的端口相同，这样，基本上所有的防火墙都不会阻止 WebSocket 连接。
- WebSocket 使用 HTTP 协议进行握手，因此它可以自然而然地集成到网络浏览器和 HTTP 服务器中，而不需要额外的成本。
- 心跳消息（ping 和 pong）将被反复的发送，进而保持 WebSocket 连接一直处于活跃状态。
- 使用该协议，当消息启动或者到达的时候，服务端和客户端都可以知道。
- WebSocket 连接关闭时将发送一个特殊的关闭消息。
- WebSocket 支持跨域，可以避免 Ajax 的限制。
- HTTP 规范要求浏览器将并发连接数限制为每个主机名两个连接，但是当我们使用 WebSocket 的时候，当握手完成之后，该限制就不存在了，因为此时的连接已经不再是 HTTP 连接了。
- WebSocket 协议支持扩展，用户可以扩展协议，实现部分自定义的子协议。
- 更好的二进制支持以及更好的压缩效果。

WebSocket 既然具有这么多优势，使用场景当然也是非常广泛的，例如：

- 在线股票网站。
- 即时聊天。
- 多人在线游戏。
- 应用集群通信。
- 系统性能实时监控。

……

在了解了这么多 WebSocket 的基本信息后，接下来看看在 Spring Boot 中如何使用 WebSocket。

11.3　Spring Boot 整合 WebSocket

Spring Boot 对 WebSocket 提供了非常友好的支持，可以方便开发者在项目中快速集成 WebSocket 功能，实现单聊或者群聊。

11.3.1 消息群发

1. 创建项目

首先创建一个 Spring Boot 项目，添加如下依赖：

```
1   <dependency>
2       <groupId>org.springframework.boot</groupId>
3       <artifactId>spring-boot-starter-websocket</artifactId>
4   </dependency>
5   <dependency>
6       <groupId>org.webjars</groupId>
7       <artifactId>webjars-locator-core</artifactId>
8   </dependency>
9   <dependency>
10      <groupId>org.webjars</groupId>
11      <artifactId>sockjs-client</artifactId>
12      <version>1.1.2</version>
13  </dependency>
14  <dependency>
15      <groupId>org.webjars</groupId>
16      <artifactId>stomp-websocket</artifactId>
17      <version>2.3.3</version>
18  </dependency>
19  <dependency>
20      <groupId>org.webjars</groupId>
21      <artifactId>jquery</artifactId>
22      <version>3.3.1</version>
23  </dependency>
```

spring-boot-starter-websocket 依赖是 Web Socket 相关依赖，其他的都是前端库，使用 jar 包的形式对这些前端库进行统一管理，使用 webjar 添加到项目中的前端库，在 Spring Boot 项目中已经默认添加了静态资源过滤，因此可以直接使用。

2. 配置 WebSocket

Spring 框架提供了基于 WebSocket 的 STOMP 支持，STOMP 是一个简单的可互操作的协议，通常被用于通过中间服务器在客户端之间进行异步消息传递。WebSocket 配置如下：

```
1   @Configuration
2   @EnableWebSocketMessageBroker
3   public class WebSocketConfig
4           implements WebSocketMessageBrokerConfigurer {
5       @Override
6       public void configureMessageBroker(MessageBrokerRegistry config) {
7           config.enableSimpleBroker("/topic");
8           config.setApplicationDestinationPrefixes("/app");
9       }
10      @Override
11      public void registerStompEndpoints(StompEndpointRegistry registry) {
12          registry.addEndpoint("/chat").withSockJS();
```

```
13        }
14    }
```

代码解释：

- 自定义类 WebSocketConfig 继承自 WebSocketMessageBrokerConfigurer 进行 WebSocket 配置，然后通过@EnableWebSocketMessageBroker 注解开启 WebSocket 消息代理。
- config.enableSimpleBroker("/topic")表示设置消息代理的前缀，即如果消息的前缀是"/topic"，就会将消息转发给消息代理（broker），再由消息代理将消息广播给当前连接的客户端。
- config.setApplicationDestinationPrefixes("/app")表示配置一个或多个前缀，通过这些前缀过滤出需要被注解方法处理的消息。例如，前缀为"/app"的 destination 可以通过@MessageMapping 注解的方法处理，而其他 destination（例如"/topic""/queue"）将被直接交给 broker 处理。
- registry.addEndpoint("/chat").withSockJS()则表示定义一个前缀为"/chat"的 endPoint，并开启 sockjs 支持，sockjs 可以解决浏览器对 WebSocket 的兼容性问题，客户端将通过这里配置的 URL 来建立 WebSocket 连接。

3. 定义 Controller

定义一个 Controller 用来实现对消息的处理，代码如下：

```
1   @Controller
2   public class GreetingController {
3       @MessageMapping("/hello")
4       @SendTo("/topic/greetings")
5       public Message greeting(Message message) throws Exception {
6           return message;
7       }
8   }
9   public class Message {
10      private String name;
11      private String content;
12      //省略 getter/setter
13  }
```

根据第 2 步的配置，@MessageMapping("/hello")注解的方法将用来接收"/app/hello"路径发送来的消息，在注解方法中对消息进行处理后，再将消息转发到@SendTo 定义的路径上，而@SendTo 路径是一个前缀为"/topic"的路径，因此该消息将被交给消息代理 broker，再由 broker 进行广播。

4. 构建聊天页面

在 resources/static 目录下创建 chat.html 页面作为聊天页面，代码如下：

```
1   <!DOCTYPE html>
2   <html lang="en">
3   <head>
4   <meta charset="UTF-8">
5   <title>群聊</title>
6   <script src="/webjars/jquery/jquery.min.js"></script>
7   <script src="/webjars/sockjs-client/sockjs.min.js"></script>
8   <script src="/webjars/stomp-websocket/stomp.min.js"></script>
```

```
9   <script src="/app.js"></script>
10  </head>
11  <body>
12  <div>
13  <label for="name">请输入用户名：</label>
14  <input type="text" id="name" placeholder="用户名">
15  </div>
16  <div>
17  <button id="connect" type="button">连接</button>
18  <button id="disconnect" type="button" disabled="disabled">断开连接</button>
19  </div>
20  <div id="chat" style="display: none;">
21  <div>
22  <label for="name">请输入聊天内容：</label>
23  <input type="text" id="content" placeholder="聊天内容">
24  </div>
25  <button id="send" type="button">发送</button>
26  <div id="greetings">
27  <div id="conversation" style="display: none">群聊进行中...</div>
28  </div>
29  </div>
30  </body>
31  </html>
```

注意，第 6~8 行是引入外部的 JS 库，这些 JS 库在 pom.xml 文件中通过依赖加入进来。app.js 是一个自定义 JS，代码如下：

```
1   var stompClient = null;
2   function setConnected(connected) {
3       $("#connect").prop("disabled", connected);
4       $("#disconnect").prop("disabled", !connected);
5       if (connected) {
6           $("#conversation").show();
7           $("#chat").show();
8       }
9       else {
10          $("#conversation").hide();
11          $("#chat").hide();
12      }
13      $("#greetings").html("");
14  }
15  function connect() {
16      if (!$("#name").val()) {
17          return;
18      }
19      var socket = new SockJS('/chat');
20      stompClient = Stomp.over(socket);
21      stompClient.connect({}, function (frame) {
22          setConnected(true);
23          stompClient.subscribe('/topic/greetings', function (greeting) {
24              showGreeting(JSON.parse(greeting.body));
```

```
25            });
26        });
27  }
28  function disconnect() {
29      if (stompClient !== null) {
30          stompClient.disconnect();
31      }
32      setConnected(false);
33  }
34  function sendName() {
35      stompClient.send("/app/hello",{},
36          JSON.stringify({'name': $("#name").val(),'content':$("#content").val()}));
37  }
38  function showGreeting(message) {
39      $("#greetings")
40          .append("<div>" + message.name+":"+message.content + "</div>");
41  }
42  $(function () {
43      $( "#connect" ).click(function() { connect(); });
44      $( "#disconnect" ).click(function() { disconnect(); });
45      $( "#send" ).click(function() { sendName(); });
46  });
```

代码解释：

- connect 方法表示建立一个 WebSocket 连接，在建立 WebSocket 连接时，用户必须先输入用户名，然后才能建立连接。
- 第 19~26 行首先使用 SockJS 建立连接，然后创建一个 STOMP 实例发起连接请求，在连接成功的回调方法中，首先调用 setConnected(true);方法进行页面的设置，然后调用 STOMP 中的 subscribe 方法订阅服务端发送回来的消息，并将服务端发送来的消息展示出来（使用 showGreeting 方法）。
- 调用 STOMP 中的 disconnect 方法可以断开一个 WebSocket 连接。

5. 测试

接下来启动 Spring Boot 项目进行测试，在浏览器中输入 http://localhost:8080/chat.html，显示结果如图 11-1 所示。

图 11-1

用户首先输入用户名，然后单击"连接"按钮，结果如图 11-2 所示。

图 11-2

然后换一个浏览器，或者使用 Chrome 浏览器的多用户（注意不是多窗口），重复刚才的步骤，这样就有两个用户连接上了，接下来就可以开始群聊了（当然也可以有更多的用户连接上来），如图 11-3 所示。

图 11-3

11.3.2　消息点对点发送

在 11.3.1 小节中介绍的消息发送使用到了 @SendTo 注解，该注解将方法处理过的消息转发到 broker，再由 broker 进行消息广播。除了 @SendTo 注解外，Spring 还提供了 SimpMessagingTemplate 类来让开发者更加灵活地发送消息，使用 SimpMessagingTemplate 可以对 11.3.1 小节的案例中的 Controller 进行如下改造：

```
@Controller
public class GreetingController {
    @Autowired
    SimpMessagingTemplate messagingTemplate;
    @MessageMapping("/hello")
    public void greeting(Message message) throws Exception {
        messagingTemplate.convertAndSend("/topic/greetings", message);
    }
}
```

改造完成后，直接运行，和 11.3.1 小节的运行结果一致。这里使用 SimpMessagingTemplate 进行消息的发送，在 Spring Boot 中，SimpMessagingTemplate 已经配置好，开发者直接注入进来即可。使用 SimpMessagingTemplate，开发者可以在任意地方发送消息到 broker，也可以发送消息给某一个用户，这就是点对点的消息发送。接下来看看如何实现消息的点对点发送（注意本案例在 11.3.1 节案例的基础上完成）。

1．添加依赖

既然是点对点发送，就应该有用户的概念，因此，首先在项目中加入 Spring Security 的依赖，代码如下：

```
1  <dependency>
2  <groupId>org.springframework.boot</groupId>
3  <artifactId>spring-boot-starter-security</artifactId>
4  </dependency>
```

2．配置 Spring Security

对 Spring Security 进行配置，添加两个用户，同时配置所有地址都认证后才能访问，代码如下：

```
1   @Configuration
2   public class WebSecurityConfig extends WebSecurityConfigurerAdapter {
3       @Bean
4       PasswordEncoder passwordEncoder() {
5           return new BCryptPasswordEncoder();
6       }
7       @Override
8       protected void configure(AuthenticationManagerBuilder auth) throws Exception {
9           auth.inMemoryAuthentication()
10              .withUser("admin")
11              .password("$2a$10$RMuFXGQ5AtH4wOvkUqyvuecpqUSeoxZYqilXzbz50dceRsga.WYiq")
12              .roles("admin")
13              .and()
14              .withUser("sang")
15              .password("$2a$10$RMuFXGQ5AtH4wOvkUqyvuecpqUSeoxZYqilXzbz50dceRsga.WYiq")
16              .roles("user");
17      }
18      @Override
19      protected void configure(HttpSecurity http) throws Exception {
20          http.authorizeRequests()
21              .anyRequest().authenticated()
22              .and()
23              .formLogin().permitAll();
24      }
25  }
```

这里就是 Spring Security 的一个常规配置，相关配置含义可以参考 10.1 节。

3．改造 WebSocket 配置

接下来对 WebSocket 配置进行改造，代码如下：

```
1   @Configuration
2   @EnableWebSocketMessageBroker
3   public class WebSocketConfig
4           implements WebSocketMessageBrokerConfigurer {
```

```
5       @Override
6       public void configureMessageBroker(MessageBrokerRegistry config) {
7           config.enableSimpleBroker("/topic","/queue");
8           config.setApplicationDestinationPrefixes("/app");
9       }
10      @Override
11      public void registerStompEndpoints(StompEndpointRegistry registry) {
12          registry.addEndpoint("/chat").withSockJS();
13      }
14  }
```

这里的修改是在 config.enableSimpleBroker("/topic");方法的基础上又增加了一个 broker 前缀 "/queue"，方便对群发消息和点对点消息进行管理。

4．配置 Controller

对 WebSocket 的 Controller 进行改造，代码如下：

```
1   @Controller
2   public class GreetingController {
3       @Autowired
4       SimpMessagingTemplate messagingTemplate;
5       @MessageMapping("/hello")
6       @SendTo("/topic/greetings")
7       public Message greeting(Message message) throws Exception {
8           return message;
9       }
10      @MessageMapping("/chat")
11      public void chat(Principal principal, Chat chat) {
12          String from = principal.getName();
13          chat.setFrom(from);
14          messagingTemplate.convertAndSendToUser(chat.getTo(),
15  "/queue/chat", chat);
16      }
17  }
18  public class Chat {
19      private String to;
20      private String from;
21      private String content;
22      //省略 getter/setter
23  }
```

代码解释：

- 群发消息依然使用@SendTo 注解来实现，点对点的消息发送则使用 SimpMessagingTemplate 来实现。
- 第 10~16 行定义了一个新的消息处理接口，@MessageMapping("/chat")注解表示来自 "/app/chat" 路径的消息将被 chat 方法处理。chat 方法的第一个参数 Principal 可以用来获取当前登录用户的信息，第二个参数则是客户端发送来的消息。
- 在 chat 方法中，首先获取当前用户的用户名，设置给 chat 对象的 from 属性，再将消息发送出去，发送的目标用户就是 chat 对象的 to 属性值。
- 消息发送使用的方法是 convertAndSendToUser，该方法内部调用了 convertAndSend 方法，并对消息路径做了处理，部分源码如下：

```java
public void convertAndSendToUser(String user, String destination, Object payload,
        @Nullable Map<String, Object> headers,
@Nullable MessagePostProcessor postProcessor)
        throws MessagingException {
  ……
……
    super.convertAndSend(this.destinationPrefix + user + destination,
payload, headers, postProcessor);
}
```

这里 destinationPrefix 的默认值是 "/user"，也就是说消息的最终发送路径是 "/user/用户名/queue/chat"。
- chat 是一个普通的 JavaBean，to 属性表示消息的目标用户，from 表示消息从哪里来，content 则是消息的主体内容。

5. 创建在线聊天页面

在 resources/static 目录下创建 onlinechat.html 页面作为在线聊天页面，代码如下：

```html
<!DOCTYPE html>
<html lang="en">
<head>
<meta charset="UTF-8">
<title>单聊</title>
<script src="/webjars/jquery/jquery.min.js"></script>
<script src="/webjars/sockjs-client/sockjs.min.js"></script>
<script src="/webjars/stomp-websocket/stomp.min.js"></script>
<script src="/chat.js"></script>
</head>
<body>
<div id="chat">
<div id="chatsContent">
</div>
<div>
        请输入聊天内容：
<input type="text" id="content" placeholder="聊天内容">
        目标用户：
<input type="text" id="to" placeholder="目标用户">
<button id="send" type="button">发送</button>
</div>
</div>
</body>
</html>
```

这个页面和 chat.html 页面基本类似，不同的是，为了演示方便，这里需要用户手动输入目标用户名。另外，还有一个 chat.js 文件，代码如下：

```
1   var stompClient = null;
2   function connect() {
3       var socket = new SockJS('/chat');
4       stompClient = Stomp.over(socket);
5       stompClient.connect({}, function (frame) {
6           stompClient.subscribe('/user/queue/chat', function (chat) {
7               showGreeting(JSON.parse(chat.body));
8           });
9       });
10  }
11  function sendMsg() {
12      stompClient.send("/app/chat", {},
13          JSON.stringify({'content':$("#content").val(),
14          'to':$("#to").val()}));
15  }
16  function showGreeting(message) {
17      $("#chatsContent")
18          .append("<div>" + message.from+":"+message.content + "</div>");
19  }
20  $(function () {
21      connect();
22      $( "#send" ).click(function() { sendMsg(); });
23  });
```

chat.js 文件基本与前文的 app.js 文件内容一致，差异主要体现在三个地方：

- 连接成功后，订阅的地址为 "/user/queue/chat"，该地址比服务端配置的地址多了 "/user" 前缀，这是因为 SimpMessagingTemplate 类中自动添加了路径前缀。
- 聊天消息发送路径为 "/app/chat"。
- 发送的消息内容中有一个 to 字段，该字段用来描述消息的目标用户。

6. 测试

经过如上几个步骤之后，一个点对点的聊天服务就搭建成功了，接下来直接在浏览器地址栏中输入 http://localhost:8080/onlinechat.html，首先会自动跳转到 Spring Security 的默认登录页面，分别使用一开始配置的两个用户 admin/123 和 sang/123 登录，登录成功后，就可以开始在线聊天了，如图 11-4 所示。

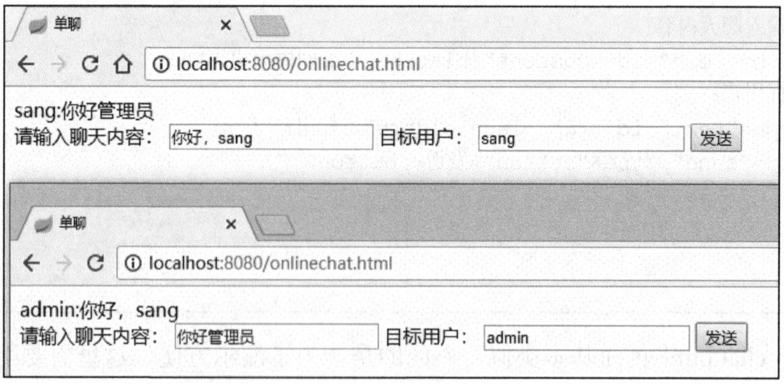

图 11-4

11.4 小　　结

本章主要向读者介绍了 Spring Boot 整合 WebSocket，整体来说，经过 Spring Boot 自动化配置之后的 WebSocket 使用起来还是非常方便的。通过@MessageMapping 注解配置消息接口，通过@SendTo 或者 SimpMessagingTemplate 进行消息转发，通过简单的几行配置，就能实现点对点、点对面的消息发送。在企业信息管理系统中，一般即时通信、通告发布等功能都会用到 WebSocket。

第 12 章

消息服务

本章概要

- JMS
- AMQP

消息队列（Message Queue）是一种进程间或者线程间的异步通信方式，使用消息队列，消息生产者在产生消息后，会将消息保存在消息队列中，直到消息消费者来取走它，即消息的发送者和接收者不需要同时与消息队列交互。使用消息队列可以有效实现服务的解耦，并提高系统的可靠性以及可扩展性。目前，开源的消息队列服务非常多，如 Apache ActiveMQ、RabbitMQ 等，这些产品也就是常说的消息中间件。

12.1 JMS

12.1.1 JMS 简介

JMS（Java Message Service）即 Java 消息服务，它通过统一 JAVA API 层面的标准，使得多个客户端可以通过 JMS 进行交互，大部分消息中间件提供商都对 JMS 提供支持。JMS 和 ActiveMQ 的关系就象 JDBC 和 JDBC 驱动的关系。JMS 包括两种消息模型：点对点和发布者/订阅者，同时 JMS 仅支持 Java 平台。

12.1.2　Spring Boot 整合 JMS

由于 JMS 是一套标准，因此 Spring Boot 整合 JMS 必然就是整合 JMS 的某一个实现，本案例以 ActiveMQ 为例来看 Spring Boot 如何进行整合。

1．ActiveMQ 简介

Apache ActiveMQ 是一个开源的消息中间件，它不仅完全支持 JMS1.1 规范，而且支持多种编程语言，例如 C、C++、C#、Delphi、Erlang、Adobe Flash、Haskell、Java、JavaScript、Perl、PHP、Pike、Python 和 Ruby 等，也支持多种协议，例如 OpenWire、REST、STOMP、WS-Notification、MQTT、XMPP 以及 AMQP。Apache ActiveMQ 也提供了对 Spring 框架的支持，可以非常容易地嵌入 Spring 中，同时它也提供了集群支持。

2．ActiveMQ 安装

一般情况下，ActiveMQ 都是安装在 Linux 上的，因此，本案例的安装环境为 CentOS 7，ActiveMQ 版本为 5.15.4，安装步骤如下（注意，要运行 ActiveMQ，CentOS 上必须安装 Java 运行环境，Java 运行环境的安装比较简单，读者可以自行安装，这里不做介绍）：

步骤01 下载 ActiveMQ，命令如下：

```
1  wget http://mirrors.hust.edu.cn/apache//activemq/5.15.4/apache-activemq-5.15.4-bin.tar.gz
```

步骤02 解压下载文件，命令如下：

```
1  Tar -zxvf apache-activemq-5.15.4-bin.tar.gz
```

步骤03 启动 ActiveMQ，命令如下：

```
1  cd apache-activemq-5.15.4
2  cd bin/
3  ./activemq start
```

步骤04 访问。

ActiveMQ 启动成功后，关闭 CentOS 防火墙，在物理机浏览器中输入地址：http://192.168.66.129:8161/，其中 192.168.66.129 是虚拟机地址，8161 是 ActiveMQ 默认端口号，如果能看到图 12-1 中的页面，表示 ActiveMQ 已经启动成功了。

图 12-1

ActiveMQ 启动成功后，单击 Manage ActiveMQ broker 超链接进入管理员控制台，默认用户名和密码都是 admin，如图 12-2 所示。

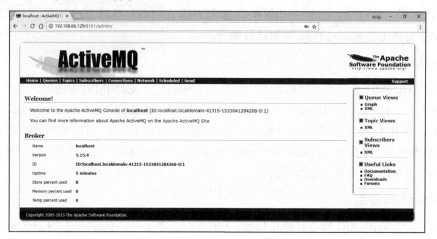

图 12-2

3. 整合 Spring Boot

Spring Boot 为 ActiveMQ 配置提供了相关的"Starter"，因此整合非常容易。

首先创建 Spring Boot 项目，添加 ActiveMQ 依赖，代码如下：

```
<dependency>
<groupId>org.springframework.boot</groupId>
<artifactId>spring-boot-starter-activemq</artifactId>
</dependency>
```

然后在 application.properties 中进行连接配置，代码如下：

```
spring.activemq.broker-url=tcp://192.168.66.129:61616
spring.activemq.packages.trust-all=true
spring.activemq.user=admin
spring.activemq.password=admin
```

首先配置 broker 地址，默认端口是 61616，然后配置信任所有的包，这个配置是为了支持发送对象消息，最后配置 ActiveMQ 的用户名和密码。

接下来在项目配置类中提供一个消息队列 Bean，该 Bean 的实例就由 ActiveMQ 提供，代码如下：

```
@Spring BootApplication
public class DemoApplication {
    public static void main(String[] args) {
        SpringApplication.run(DemoApplication.class, args);
    }
    @Bean
    Queue queue() {
        return new ActiveMQQueue("amq");
    }
}
```

接下来创建一个 JMS 组件来完成消息的发送和接收，代码如下：

```
1   @Component
2   public class JmsComponent {
3       @Autowired
4       JmsMessagingTemplate messagingTemplate;
5       @Autowired
6       Queue queue;
7       public void send(Message msg) {
8           messagingTemplate.convertAndSend(this.queue, msg);
9       }
10      @JmsListener(destination = "amq")
11      public void receive(Message msg) {
12          System.out.println("receive:" + msg);
13      }
14  }
15  public class Message implements Serializable{
16      private String content;
17      private Date date;
18      //省略 getter/setter
19  }
```

JmsMessagingTemplate 是由 Spring 提供的一个 JMS 消息发送模板，可以用来方便地进行消息的发送，消息发送方法 convertAndSend 的第一个参数是消息队列，第二个参数是消息内容，本案例演示一个对象消息。@JmsListener 注解则表示相应的方法是一个消息消费者，消息消费者订阅的消息 destination 为 amq。

经过上面的配置，就可以在 Spring Boot 中使用 ActiveMQ 了。

4．测试

编写测试类，完成消息发送测试，代码如下：

```
1   @RunWith(SpringRunner.class)
2   @SpringBootTest
3   public class DemoApplicationTests {
4       @Autowired
5       JmsComponent jmsComponent;
6       @Test
7       public void contextLoads() {
8           Message msg = new Message();
9           msg.setContent("hello jms!");
10          msg.setDate(new Date());
11          jmsComponent.send(msg);
12      }
13  }
```

在测试类中注入 JmsComponent 组件，然后调用该组件的 send 方法发送一个 Message 对象。

确认 ActiveMQ 已经启动，然后启动 Spring Boot，Spring Boot 项目启动成功后，执行该单元测试方法，观察 Spring Boot 项目日志，如图 12-3 所示。

```
2018-08-01 08:48:01.241  INFO 11088 --- [           main] com.example.demo
receive:Message{content='hello jms!', date=Wed Aug 01 08:48:16 CST 2018}
```

图 12-3

12.2　AMQP

12.2.1　AMQP 简介

AMQP（Advanced Message Queuing Protocol，高级消息队列协议）是一个线路层的协议规范，而不是 API 规范（例如 JMS）。由于 AMQP 是一个线路层协议规范，因此它天然就是跨平台的，就像 SMTP、HTTP 等协议一样，只要开发者按照规范的格式发送数据，任何平台都可以通过 AMQP 进行消息交互。像目前流行的 StormMQ、RabbitMQ 等都实现了 AMQP。

12.2.2　Spring Boot 整合 AMQP

和 JMS 一样，使用 AMQP 也是使用 AMQP 的某个实现，本案例以 RabbitMQ 为例向读者介绍 AMQP 的使用。

1．RabbitMQ 简介

RabbitMQ 是一个实现了 AMQP 的开源消息中间件，使用高性能的 Erlang 编写。RabbitMQ 具有可靠性、支持多种协议、高可用、支持消息集群以及多语言客户端等特点，在分布式系统中存储转发消息，具有不错的性能表现。

2．RabbitMQ 的安装

由于 RabbitMQ 使用 Erlang 编写，因此需要先安装 Erlang 环境，在 CentOS 7 中安装 Erlang 21.0 的步骤如下：

```
1   #下载安装包
2   wget http://erlang.org/download/otp_src_21.0.tar.gz
3   #解压文件
4   tar -zxvf otp_src_21.0.tar.gz
5   cd otp_src_21.0
6   #编译
7   ./otp_build autoconf
8   ./configure
9   make
10  #安装
11  make install
12  #检验
13  erl
```

最后一步是检验，如果看到如图 12-4 所示的效果图，表示安装成功。

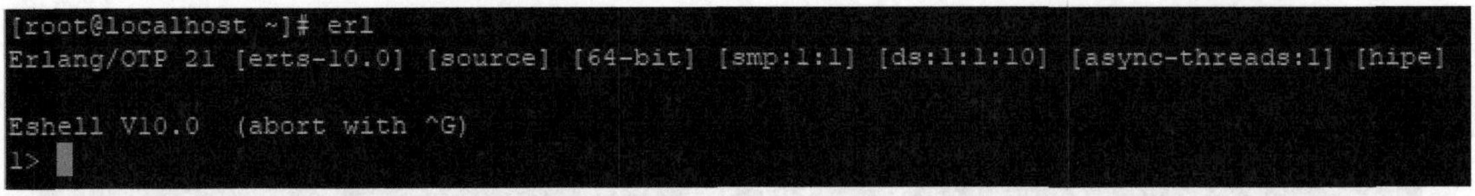

```
[root@localhost ~]# erl
Erlang/OTP 21 [erts-10.0] [source] [64-bit] [smp:1:1] [ds:1:1:10] [async-threads:1] [hipe]

Eshell V10.0  (abort with ^G)
1>
```

图 12-4

Erlang 安装成功后，接下来安装 RabbitMQ。

由于 yum 仓库中默认的 Erlang 版本较低，因此首先需要将最新的 Erlang 包添加到 yum 源中，执行如下命令：

| 1 | vi /etc/yum.repos.d/rabbitmq-erlang.repo |

添加如下内容：

1	[rabbitmq-erlang]
2	name=rabbitmq-erlang
3	baseurl=https://dl.bintray.com/rabbitmq/rpm/erlang/21/el/7
4	gpgcheck=1
5	gpgkey=https://dl.bintray.com/rabbitmq/Keys/rabbitmq-release-signing-key.asc
6	repo_gpgcheck=0
7	enabled=1

添加成功后，清除原有缓存并创建新缓存，命令如下：

1	yum clean all
2	yum makecache

准备工作完成后，接下来就可以安装 RabbitMQ 了，首先下载文件：

| 1 | wget https://dl.bintray.com/rabbitmq/all/rabbitmq-server/3.7.7/rabbitmq-server-3.7.7-1.el7.noarch.rpm |

下载完成后开始安装：

| 1 | yum install rabbitmq-server-3.7.7-1.el7.noarch.rpm |

安装过程中，若提示缺少 socat 依赖，则安装 socat 依赖即可，命令如下：

| 1 | yum install socat |

安装成功后，接下来就可以启动 RabbitMQ 并进行用户管理了，命令如下：

1	#启动
2	service rabbitmq-server start
3	#查看状态
4	rabbitmqctl status
5	#开启 web 插件
6	rabbitmq-plugins enable rabbitmq_management
7	#重启
8	service rabbitmq-server restart
9	#添加一个用户名为 sang，密码为 123 的用户
10	rabbitmqctl add_user sang 123
11	#设置 sang 用户的角色为管理员

```
12  rabbitmqctl set_user_tags sang administrator
13  #配置 sang 用户可以远程登录
14  rabbitmqctl set_permissions -p / sang ".*" ".*" ".*"
```

> **注 意**
>
> RabbitMQ 启动成功后，默认有一个 guest 用户，但是该用户只能在本地登录，无法远程登录，因此本案例中添加了一个新的用户 sang，也具有管理员身份，同时可以远程登录。当 RabbitMQ 启动成功后，在物理机浏览器上输入虚拟机地址：http://192.168.66.130:15672，如果看到如图 12-5 所示的页面，表示 RabbitMQ 已经启动成功。使用 sang/123 进行登录，登录成功后如图 12-6 所示。

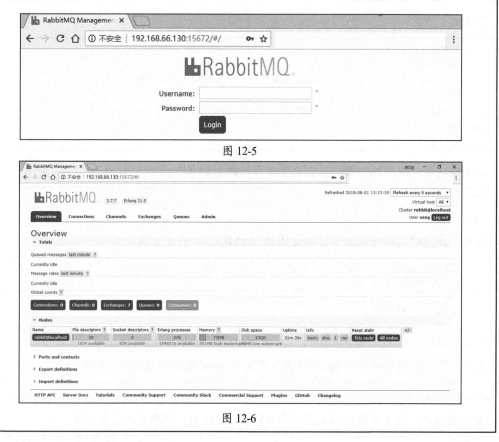

图 12-5

图 12-6

3. 整合 Spring Boot

Spring Boot 为 AMQP 提供了自动化配置依赖 spring-boot-starter-amqp，因此首先创建 Spring Boot 项目并添加该依赖，代码如下：

```
1  <dependency>
2      <groupId>org.springframework.boot</groupId>
3      <artifactId>spring-boot-starter-amqp</artifactId>
4  </dependency>
```

项目创建成功后，在 application.properties 中配置 RabbitMQ 的基本连接信息，代码如下：

```
1  spring.rabbitmq.host=192.168.66.130
2  spring.rabbitmq.port=5672
3  spring.rabbitmq.username=sang
4  spring.rabbitmq.password=123
```

接下来进行 RabbitMQ 配置，在 RabbitMQ 中，所有的消息生产者提交的消息都会交由 Exchange 进行再分配，Exchange 会根据不同的策略将消息分发到不同的 Queue 中。RabbitMQ 中一共提供了 4 种不同的 Exchange 策略，分别是 Direct、Fanout、Topic 以及 Header，这 4 种不同的策略中，前 3 种的使用频率较高，第 4 种的使用频率较低，下面分别对这 4 种不同的 ExchangeType 予以介绍。

（1）Direct

DirectExchange 的路由策略是将消息队列绑定到一个 DirectExchange 上，当一条消息到达 DirectExchange 时会被转发到与该条消息 routing key 相同的 Queue 上，例如消息队列名为 "hello-queue"，则 routingkey 为 "hello-queue" 的消息会被该消息队列接收。DirectExchange 的配置如下：

```
1   @Configuration
2   public class RabbitDirectConfig {
3       public final static String DIRECTNAME = "sang-direct";
4       @Bean
5       Queue queue() {
6           return new Queue("hello-queue");
7       }
8       @Bean
9   DirectExchange directExchange() {
10          return new DirectExchange(DIRECTNAME, true, false);
11      }
12      @Bean
13      Binding binding() {
14          return BindingBuilder.bind(queue())
15              .to(directExchange()).with("direct");
16      }
17  }
```

代码解释：

- 首先提供一个消息队列 Queue，然后创建一个 DirectExchange 对象，三个参数分别是名字、重启后是否依然有效以及长期未用时是否删除。
- 创建一个 Binding 对象，将 Exchange 和 Queue 绑定在一起。
- DirectExchange 和 Binding 两个 Bean 的配置可以省略掉，即如果使用 DirectExchange，只配置一个 Queue 的实例即可。

接下来配置一个消费者，代码如下：

```
1   @Component
2   public class DirectReceiver {
3       @RabbitListener(queues = "hello-queue")
4       public void handler1(String msg) {
5           System.out.println("DirectReceiver:" + msg);
```

```
6       }
7   }
```

通过@RabbitListener 注解指定一个方法是一个消息消费方法，方法参数就是所接收到的消息。然后在单元测试类中注入一个 RabbitTemplate 对象来进行消息发送，代码如下：

```
1   @RunWith(SpringRunner.class)
2   @SpringBootTest
3   public class RabbitmqApplicationTests {
4       @Autowired
5       RabbitTemplate rabbitTemplate;
6       @Test
7       public void directTest() {
8           rabbitTemplate.convertAndSend("hello-queue", "hello direct!");
9       }
10  }
```

确认 RabbitMQ 已经启动，然后启动 Spring Boot 项目，启动成功后，运行该单元测试方法，在 Spring Boot 控制台打印日志，如图 12-7 所示。

```
2018-08-02 10:26:05.190  INFO 11696 --- [
DirectReceiver:hello direct!
```

图 12-7

（2）Fanout

FanoutExchange 的数据交换策略是把所有到达 FanoutExchange 的消息转发给所有与它绑定的 Queue，在这种策略中，routingkey 将不起任何作用，FanoutExchange 的配置方式如下：

```
1   @Configuration
2   public class RabbitFanoutConfig {
3       public final static String FANOUTNAME = "sang-fanout";
4       @Bean
5       FanoutExchange fanoutExchange() {
6           return new FanoutExchange(FANOUTNAME, true, false);
7       }
8       @Bean
9       Queue queueOne() {
10          return new Queue("queue-one");
11      }
12      @Bean
13      Queue queueTwo() {
14          return new Queue("queue-two");
15      }
16      @Bean
17      Binding bindingOne() {
18          return BindingBuilder.bind(queueOne()).to(fanoutExchange());
19      }
20      @Bean
21      Binding bindingTwo() {
```

```
22          return BindingBuilder.bind(queueTwo()).to(fanoutExchange());
23      }
24  }
```

在这里首先创建 FanoutExchange，参数的含义与创建 DirectExchange 参数的含义一致，然后创建两个 Queue，再将这两个 Queue 都绑定到 FanoutExchange 上。接下来创建两个消费者，代码如下：

```
1   @Component
2   public class FanoutReceiver {
3       @RabbitListener(queues = "queue-one")
4       public void handler1(String message) {
5           System.out.println("FanoutReceiver:handler1:" + message);
6       }
7       @RabbitListener(queues = "queue-two")
8       public void handler2(String message) {
9           System.out.println("FanoutReceiver:handler2:" + message);
10      }
11  }
```

两个消费者分别消费两个消息队列中的消息，然后在单元测试中发送消息，代码如下：

```
1   @RunWith(SpringRunner.class)
2   @Spring BootTest
3   public class RabbitmqApplicationTests {
4       @Autowired
5       RabbitTemplate rabbitTemplate;
6       @Test
7       public void fanoutTest() {
8           rabbitTemplate
9               .convertAndSend(RabbitFanoutConfig.FANOUTNAME,
10                  null, "hello fanout!");
11      }
12  }
```

注意，这里发送消息时不需要 routingkey，指定 exchange 即可，routingkey 可以直接传一个 null。

确认 RabbitMQ 已经启动，然后启动 Spring Boot 项目，启动成功后，执行单元测试方法，控制台打印日志如图 12-8 所示。

```
2018-08-02 11:21:55.654  INFO 11804 --- [
FanoutReceiver:handler2:hello fanout!
FanoutReceiver:handler1:hello fanout!
```

图 12-8

可以看到，一条消息发送出去之后，所有和该 FanoutExchange 绑定的 Queue 都收到了消息。

（3）Topic

TopicExchange 是比较复杂也比较灵活的一种路由策略，在 TopicExchange 中，Queue 通过 routingkey 绑定到 TopicExchange 上，当消息到达 TopicExchange 后，TopicExchange 根据消息的

routingkey 将消息路由到一个或者多个 Queue 上。TopicExchange 配置如下：

```
1    @Configuration
2    public class RabbitTopicConfig {
3        public final static String TOPICNAME = "sang-topic";
4        @Bean
5        TopicExchange topicExchange() {
6            return new TopicExchange(TOPICNAME, true, false);
7        }
8        @Bean
9        Queue xiaomi() {
10           return new Queue("xiaomi");
11       }
12       @Bean
13       Queue huawei() {
14           return new Queue("huawei");
15       }
16       @Bean
17       Queue phone() {
18           return new Queue("phone");
19       }
20       @Bean
21       Binding xiaomiBinding() {
22           return BindingBuilder.bind(xiaomi()).to(topicExchange())
23                   .with("xiaomi.#");
24       }
25       @Bean
26       Binding huaweiBinding() {
27           return BindingBuilder.bind(huawei()).to(topicExchange())
28                   .with("huawei.#");
29       }
30       @Bean
31       Binding phoneBinding() {
32           return BindingBuilder.bind(phone()).to(topicExchange())
33                   .with("#.phone.#");
34       }
35   }
```

代码解释：

- 首先创建 TopicExchange，参数和前面的一致。然后创建三个 Queue，第一个 Queue 用来存储和"xiaomi"有关的消息，第二个 Queue 用来存储和"huawei"有关的消息，第三个 Queue 用来存储和"phone"有关的消息。
- 将三个 Queue 分别绑定到 TopicExchange 上，第一个 Binding 中的"xiaomi.#"表示消息的 routingkey 凡是以"xiaomi"开头的，都将被路由到名称为"xiaomi"的 Queue 上；第二个 Binding 中的"huawei.#"表示消息的 routingkey 凡是以"huawei"开头的，都将被路由到名称为"huawei"的 Queue 上；第三个 Binding 中的"#.phone.#"则表示消息的 routingkey 中凡是包含"phone"的，都将被路由到名称为"phone"的 Queue 上。

接下来针对三个 Queue 创建三个消费者，代码如下：

```
1  @Component
2  public class TopicReceiver {
3      @RabbitListener(queues = "phone")
4      public void handler1(String message) {
5          System.out.println("PhoneReceiver:" + message);
6      }
7      @RabbitListener(queues = "xiaomi")
8      public void handler2(String message) {
9          System.out.println("XiaoMiReceiver:"+message);
10     }
11     @RabbitListener(queues = "huawei")
12     public void handler3(String message) {
13         System.out.println("HuaWeiReceiver:"+message);
14     }
15 }
```

然后在单元测试中进行消息的发送，代码如下：

```
1  @RunWith(SpringRunner.class)
2  @SpringBootTest
3  public class RabbitmqApplicationTests {
4      @Autowired
5      RabbitTemplate rabbitTemplate;
6      @Test
7      public void topicTest() {
8          rabbitTemplate.convertAndSend(RabbitTopicConfig.TOPICNAME,
9  "xiaomi.news","小米新闻..");
10         rabbitTemplate.convertAndSend(RabbitTopicConfig.TOPICNAME,
11 "huawei.news","华为新闻..");
12         rabbitTemplate.convertAndSend(RabbitTopicConfig.TOPICNAME,
13 "xiaomi.phone","小米手机..");
14         rabbitTemplate.convertAndSend(RabbitTopicConfig.TOPICNAME,
15 "huawei.phone","华为手机..");
16         rabbitTemplate.convertAndSend(RabbitTopicConfig.TOPICNAME,
17 "phone.news","手机新闻..");
18     }
19 }
```

根据 RabbitTopicConfig 中的配置，第一条消息将被路由到名称为"xiaomi"的 Queue 上，第二条消息将被路由到名为"huawei"的 Queue 上，第三条消息将被路由到名为"xiaomi"以及名为"phone"的 Queue 上，第四条消息将被路由到名为"huawei"以及名为"phone"的 Queue 上，最后一条消息则将被路由到名为"phone"的 Queue 上。

确认 RabbitMQ 已经启动，然后启动 Spring Boot 项目，启动成功后，运行单元测试方法，控制台打印日志如图 12-9、图 12-10 所示。

```
2018-08-02 12:13:12.419  INFO 5700 --- [
HuaWeiReceiver:华为新闻..
XiaoMiReceiver:小米新闻..
PhoneReceiver:小米手机..
PhoneReceiver:手机新闻..
```

图 12-9

```
2018-08-02 12:13:21.928  INFO 10532 --- [
HuaWeiReceiver:华为手机..
XiaoMiReceiver:小米手机..
PhoneReceiver:华为手机..
```

图 12-10

（4）Header

HeadersExchange 是一种使用较少的路由策略，HeadersExchange 会根据消息的 Header 将消息路由到不同的 Queue 上，这种策略也和 routingkey 无关，配置如下：

```
@Configuration
public class RabbitHeaderConfig {
    public final static String HEADERNAME = "sang-header";
    @Bean
    HeadersExchange headersExchange() {
        return new HeadersExchange(HEADERNAME, true, false);
    }
    @Bean
    Queue queueName() {
        return new Queue("name-queue");
    }
    @Bean
    Queue queueAge() {
        return new Queue("age-queue");
    }
    @Bean
    Binding bindingName() {
        Map<String, Object> map = new HashMap<>();
        map.put("name", "sang");
        return BindingBuilder.bind(queueName())
                .to(headersExchange()).whereAny(map).match();
    }
    @Bean
    Binding bindingAge() {
        return BindingBuilder.bind(queueAge())
                .to(headersExchange()).where("age").exists();
    }
}
```

这里的配置大部分和前面介绍的一样，差别主要体现的 Binding 的配置上，第一个 bindingName

方法中，whereAny 表示消息的 Header 中只要有一个 Header 匹配上 map 中的 key/value，就把该消息路由到名为 "name-queue" 的 Queue 上，这里也可以使用 whereAll 方法，表示消息的所有 Header 都要匹配。whereAny 和 whereAll 实际上对应了一个名为 x-match 的属性。bindingAge 中的配置则表示只要消息的 Header 中包含 age，无论 age 的值是多少，都将消息路由到名为 "age-queue" 的 Queue 上。

接下来创建两个消息消费者，代码如下：

```
1   @Component
2   public class HeaderReceiver {
3       @RabbitListener(queues = "name-queue")
4       public void handler1(byte[] msg) {
5           System.out.println("HeaderReceiver:name:"
6                   + new String(msg, 0, msg.length));
7       }
8       @RabbitListener(queues = "age-queue")
9       public void handler2(byte[] msg) {
10          System.out.println("HeaderReceiver:age:"
11                  + new String(msg, 0, msg.length));
12      }
13  }
```

注意，这里的参数用 byte 数组接收。然后在单元测试中创建消息的发送方法，这里消息的发送也和 routingkey 无关，代码如下：

```
1   @RunWith(SpringRunner.class)
2   @SpringBootTest
3   public class RabbitmqApplicationTests {
4       @Autowired
5       RabbitTemplate rabbitTemplate;
6       @Test
7       public void headerTest() {
8           Message nameMsg = MessageBuilder
9                   .withBody("hello header! name-queue".getBytes())
10                  .setHeader("name", "sang").build();
11          Message ageMsg = MessageBuilder
12                  .withBody("hello header! age-queue".getBytes())
13                  .setHeader("age", "99").build();
14          rabbitTemplate.send(RabbitHeaderConfig.HEADERNAME, null, ageMsg);
15          rabbitTemplate.send(RabbitHeaderConfig.HEADERNAME, null, nameMsg);
16      }
17  }
```

这里创建两条消息，两条消息具有不同的 header，不同 header 的消息将被发送到不同的 Queue 中。

确认 RabbitMQ 已经启动，然后启动 Spring Boot 项目，启动成功后，执行单元测试方法，结果如图 12-11 所示。

```
2018-08-02 12:37:55.730  INFO 660 --- [
HeaderReceiver:name:hello header! name-queue
HeaderReceiver:age:hello header! age-queue
```

图 12-11

12.3 小　　结

　　本章向读者介绍了 Spring Boot 对消息服务的支持，传统的 JMS 和 AMQP 各有千秋，JMS 从 API 的层面对消息中间件进行了统一，AMQP 从协议层面来统一，JMS 不支持跨平台，而 AMQP 天然地具备跨平台功能。AMQP 支持的消息模型也更加丰富，除了本章介绍的 ActiveMQ 和 RabbitMQ 之外，Spring Boot 也能方便地整合 Kafka、Artemis 等，开发者可根据实际情况选择合适的消息中间件。

第 13 章

企业开发

本章概要

- 邮件发送
- 定时任务
- 批处理
- Swagger2
- 数据校验

13.1 邮件发送

邮件发送是一个非常常见的功能,注册时的身份认证、重要通知发送等都会用到邮件发送。Sun 公司提供了 JavaMail 用来实现邮件发送,但是配置烦琐,Spring 中提供了 JavaMailSender 用来简化邮件配置,Spring Boot 则提供了 MailSenderAutoConfiguration 对邮件的发送做了进一步简化。本节就来看看 Spring Boot 中如何发送邮件。

13.1.1 发送前的准备

本节以 QQ 邮箱为例向读者介绍邮件的发送过程。使用 QQ 邮箱发送邮件,首先要申请开通POP3/SMTP 服务或者 IMAP/SMTP 服务。SMTP 全称为 Simple Mail Transfer Protocol,译作简单邮件传输协议,它定义了邮件客户端软件与 SMTP 服务器之间,以及 SMTP 服务器与 SMTP 服务器之间的通信规则。也就是说,aaa@qq.com 用户先将邮件投递到腾讯的 SMTP 服务器,这个过程就使用了 SMTP 协议,然后腾讯的 SMTP 服务器将邮件投递到网易的 SMTP 服务器,这个过程依然使用了 SMTP 协议，SMTP 服务器就是用来接收邮件的。而 POP3 全称为 Post Office Protocol3，译作邮局协议，它定义了邮件客户端与 POP3 服务器之间的通信规则。该协议在什么场景下会用到呢？当邮件到达网易的 SMTP 服务器之后，111@163.com 用户需要登录服务器查看邮件，这个时候就用上该协议了：邮件服务商会为每一个用户提供专门的邮件存储空间，SMTP 服务器收到邮件之后，将邮件保存到相应用户的邮件存储空间中，如果用户要读取邮件，就需要通过邮件服务商的 POP3 邮件服务器来完成。至于 IMAP 协议，则是对 POP3 协议的扩展，功能更强，作用类似。下面介绍 QQ 邮箱开通 POP3/SMTP 服务或者 IMAP/SMTP 服务的步骤。

步骤01 登录 QQ 邮箱，依次单击顶部的设置按钮（见图 13-1）和账户按钮（见图 13-2）。

图 13-1

图 13-2

步骤02 在账户选项卡下方找到 POP3/SMTP 服务，单击后方的"开启"按钮，如图 13-3 所示。

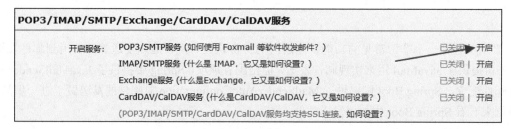

图 13-3

单击"开启"按钮后，按照引导步骤发送短信，操作成功后，会获取一个授权码（见图 13-4），将授权码保存下来过后使用。

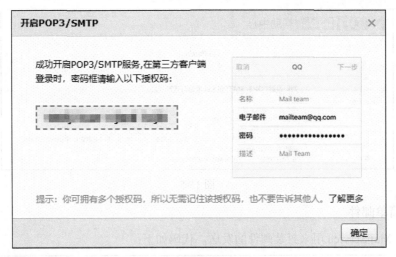

图 13-4

拿到授权码后，准备工作就完成了。

13.1.2 发送

1. 环境搭建

使用 Spring Boot 发送邮件，环境搭建非常容易，首先在创建项目时添加邮件依赖，代码如下：

```
1  <dependency>
2    <groupId>org.springframework.boot</groupId>
3    <artifactId>spring-boot-starter-mail</artifactId>
4  </dependency>
```

项目创建成功后，在 application.properties 中完成邮件基本信息配置，代码如下：

```
1  spring.mail.host=smtp.qq.com
2  spring.mail.port=465
3  spring.mail.username=xxx@qq.com
4  spring.mail.password=13.1.1 小节申请到的授权码
5  spring.mail.default-encoding=UTF-8
6  spring.mail.properties.mail.smtp.socketFactory.class=javax.net.ssl.SSLSocketFactory
7  spring.mail.properties.mail.debug=true
```

这里配置了邮件服务器的地址、端口（可以是 465 或者 587）、用户的账号和密码以及默认编码、SSL 连接配置等，最后开启 debug，这样方便开发者查看邮件发送日志。注意，SSL 的配置可以在 QQ 邮箱帮助中心看到相关文档，如图 13-5 所示。

完成这些配置之后，基本的邮件发送环境就搭建成功了，接下来就可以发送邮件了。邮件从简单到复杂有多种类型，下面分别予以介绍。

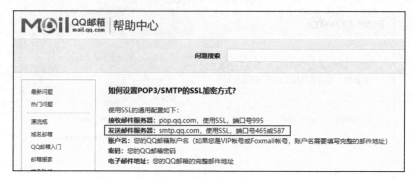

图 13-5

2. 发送简单邮件

创建一个 MailService 用来封装邮件的发送，代码如下：

```
1   @Component
2   public class MailService {
3       @Autowired
4       JavaMailSender javaMailSender;
5       public void sendSimpleMail(String from,String to,String cc,
6                       String subject,String content) {
7           SimpleMailMessage simpMsg = new SimpleMailMessage();
8           simpMsg.setFrom(from);
9           simpMsg.setTo(to);
10          simpMsg.setCc(cc);
11          simpMsg.setSubject(subject);
12          simpMsg.setText(content);
13          javaMailSender.send(simpMsg);
14      }
15  }
```

代码解释：

- JavaMailSender 是 Spring Boot 在 MailSenderPropertiesConfiguration 类中配置好的，该类在 Mail 自动配置类 MailSenderAutoConfiguration 中导入，因此这里注入 JavaMailSender 就可以使用了。
- sendSimpleMail 方法的 5 个参数分别表示邮件发送者、收件人、抄送人、邮件主题以及邮件内容。
- 简单邮件可以直接构建一个 SimpleMailMessage 对象进行配置，配置完成后，通过 JavaMailSender 将邮件发送出去。

配置完成后，在单元测试中写一个测试方法进行测试，代码如下：

```
1   @RunWith(SpringRunner.class)
2   @Spring BootTest
3   public class SendmailApplicationTests {
4       @Autowired
5       MailService mailService;
6       @Test
```

```
7       public void sendSimpleMail() {
8           mailService.sendSimpleMail("1510161612@qq.com",
9   "584991843@qq.com",
10  "1470249098@qq.com",
11  "测试邮件主题",
12  "测试邮件内容");
13      }
14  }
```

执行该方法，即可看到邮件发送成功，如图 13-6 所示。

图 13-6

3．发送带附件的邮件

要发送一个带附件的邮件也非常容易，通过调用 Attachment 方法即可添加附件，该方法调用多次即可添加多个附件。在 MailService 中添加如下方法：

```
1   public void sendAttachFileMail(String from, String to,
2               String subject, String content, File file) {
3       try {
4           MimeMessage message = javaMailSender.createMimeMessage();
5           MimeMessageHelper helper = new MimeMessageHelper(message,true);
6           helper.setFrom(from);
7           helper.setTo(to);
8           helper.setSubject(subject);
9           helper.setText(content);
10          helper.addAttachment(file.getName(), file);
11          javaMailSender.send(message);
12      } catch (MessagingException e) {
13          e.printStackTrace();
14      }
15  }
```

这里使用 MimeMessageHelper 简化了邮件配置，它的构造方法的第二个参数 true 表示构造一个 multipart message 类型的邮件，multipart message 类型的邮件包含多个正文、附件以及内嵌资源，邮件的表现形式更加丰富。最后通过 addAttachment 方法添加附件。

在单元测试中添加如下方法进行测试：

```
1   @Test
2   public void sendAttachFileMail() {
3       mailService.sendAttachFileMail("1510161612@qq.com",
4   "584991843@qq.com",
5   "测试邮件主题",
6   "测试邮件内容",
7               new File("E:\\邮件附件.docx"));
8   }
```

执行单元测试方法，邮件发送结果如图 13-7 所示。

图 13-7

4．发送带图片资源的邮件

有的邮件正文中可能要插入图片，使用 FileSystemResource 可以实现这一功能，代码如下：

```
1   public void sendMailWithImg(String from, String to,
2                       String subject, String content,
3                       String[] srcPath,String[] resIds) {
4       if (srcPath.length != resIds.length) {
5           System.out.println("发送失败");
6           return;
7       }
8       try {
9           MimeMessage message = javaMailSender.createMimeMessage();
10          MimeMessageHelper helper = new MimeMessageHelper(message,true);
11          helper.setFrom(from);
12          helper.setTo(to);
13          helper.setSubject(subject);
14          helper.setText(content,true);
15          for (int i = 0; i < srcPath.length; i++) {
16              FileSystemResource res = new FileSystemResource(new File(srcPath[i]));
```

```
17          helper.addInline(resIds[i], res);
18      }
19      javaMailSender.send(message);
20  } catch (MessagingException e) {
21      System.out.println("发送失败");
22  }
23 }
```

在发送邮件时分别传入图片资源路径和资源 id，通过 FileSystemResource 构造静态资源，然后调用 addInline 方法将资源加入邮件对象中。注意，在调用 MimeMessageHelper 中的 setText 方法时，第二个参数 true 表示邮件正文是 HTML 格式的，该参数不传默认为 false。

接下来在测试类中添加如下方法进行测试：

```
1  @Test
2  public void sendMailWithImg() {
3      mailService.sendMailWithImg("1510161612@qq.com",
4  "584991843@qq.com",
5  "测试邮件主题(图片)",
6  "<div>hello,这是一封带图片资源的邮件：" +
7  "这是图片 1: <div><img src='cid:p01'/></div>" +
8  "这是图片 2: <div><img src='cid:p02'/></div>" +
9  "</div>",
10         new String[]{"C:\\Users\\sang\\Pictures\\p1.png",
11 "C:\\Users\\sang\\Pictures\\p2.png"},
12         new String[]{"p01","p02"});
13 }
```

邮件的正文是一段 HTML 文本，用 cid 标注出两个静态资源，分别为 p01 和 p02。执行该方法，邮件发送结果如图 13-8 所示。

图 13-8

5. 使用 FreeMarker 构建邮件模板

对于格式复杂的邮件，如果采用字符串进行 HTML 拼接，不但容易出错，而且不易于维护，使用 HTML 模板可以很好地解决这一问题。使用 FreeMarker 构建邮件模板，首先加入 FreeMarker 依赖，代码如下：

```
1  <dependency>
2      <groupId>org.springframework.boot</groupId>
3      <artifactId>spring-boot-starter-freemarker</artifactId>
4  </dependency>
```

然后在 MailService 中添加如下方法：

```
1   public void sendHtmlMail(String from, String to,
2                   String subject, String content){
3       try {
4           MimeMessage message = javaMailSender.createMimeMessage();
5           MimeMessageHelper helper = new MimeMessageHelper(message, true);
6           helper.setTo(to);
7           helper.setFrom(from);
8           helper.setSubject(subject);
9           helper.setText(content, true);
10          javaMailSender.send(message);
11      } catch (MessagingException e) {
12          System.out.println("发送失败");
13      }
14  }
```

接下来在 resources 目录下创建 ftl 目录作为模板存放位置，在该目录下创建 mailtemplate.ftl 作为邮件模板，内容如下：

```
1   <div>邮箱激活</div>
2   <div>您的注册信息是:
3   <table border="1">
4   <tr>
5   <td>用户名</td>
6   <td>${username}</td>
7   </tr>
8   <tr>
9   <td>用户性别</td>
10  <td>${gender}</td>
11  </tr>
12  </table>
13  </div>
14  <div>
15  <a href="http://www.baidu.com">核对无误请点击本链接激活邮箱</a>
16  </div>
```

当然，再创建一个简单的 User 实体类，代码如下：

```
1   public class User {
2       private String username;
```

```
3       private String gender;
4       //省略getter/setter
5   }
```

最后，在单元测试类中添加如下方法进行测试：

```
1   @Test
2   public void sendHtmlMail(){
3       try {
4           Configuration configuration =
5                   new Configuration(Configuration.VERSION_2_3_0);
6           ClassLoader loader = SendmailApplication.class.getClassLoader();
7           configuration
8           .setClassLoaderForTemplateLoading(loader,"ftl");
9           Template template = configuration.getTemplate("mailtemplate.ftl");
10          StringWriter mail = new StringWriter();
11          User user = new User();
12          user.setGender("男");
13          user.setUsername("江南一点雨");
14          template.process(user, mail);
15          mailService.sendHtmlMail("1510161612@qq.com",
16  "584991843@qq.com",
17  "测试邮件主题",
18                  mail.toString());
19      } catch (Exception e) {
20          e.printStackTrace();
21      }
22  }
```

首先配置 FreeMarker 模板位置，配置模板文件，然后结合 User 对象渲染模板，将渲染结果发送出去，执行该方法，邮件发送结果如图 13-9 所示。

图 13-9

6. 使用 Thymeleaf 构建邮件模板

既然可以使用 FreeMarker 构建邮件模板，当然也可以使用 Thymeleaf 构建邮件模板，使用 Thymeleaf 构建邮件模板相对来说更加方便。使用 Thymeleaf 构建邮件模板，首先添加 Thymeleaf 依赖，代码如下：

```
1   <dependency>
```

```
2       <groupId>org.springframework.boot</groupId>
3       <artifactId>spring-boot-starter-thymeleaf</artifactId>
4   </dependency>
```

Thymeleaf 邮件模板默认位置是在 resources/templates 目录下，创建相应的目录，然后创建邮件模板 mailtemplate.html，代码如下：

```
1   <html lang="en" xmlns:th="http://www.thymeleaf.org">
2   <head>
3   <meta charset="UTF-8">
4   <title>邮件</title>
5   </head>
6   <body>
7   <div>邮箱激活</div>
8   <div>您的注册信息是：
9   <table border="1">
10  <tr>
11  <td>用户名</td>
12  <td th:text="${username}"></td>
13  </tr>
14  <tr>
15  <td>用户性别</td>
16  <td th:text="${gender}"></td>
17  </tr>
18  </table>
19  </div>
20  <div>
21  <a href="http://www.baidu.com">核对无误请点击本链接激活邮箱</a>
22  </div>
23  </body>
24  </html>
```

然后在单元测试类中添加如下代码进行测试：

```
1   @Autowired
2   TemplateEngine templateEngine;
3   @Test
4   public void sendHtmlMailThymeleaf() {
5       Context ctx = new Context();
6       ctx.setVariable("username", "sang");
7       ctx.setVariable("gender", "男");
8       String mail = templateEngine.process("mailtemplate.html", ctx);
9       mailService.sendHtmlMail("1510161612@qq.com",
10  "584991843@qq.com",
11  "测试邮件主题",
12          mail);
13  }
```

不同于 FreeMarker，Thymeleaf 提供了 TemplateEngine 来对模板进行渲染，通过 Context 构造模板中变量需要的值，这种方式比 FreeMarker 构建邮件模板更加方便。最后执行该测试方法，发送的邮件如图 13-10 所示。

图 13-10

这几种不同的邮件发送方式基本上能满足大部分的业务需求，读者在实际开发中可以合理选择。

13.2 定时任务

定时任务是企业级开发中最常见的功能之一，如定时统计订单数、数据库备份、定时发送短信和邮件、定时统计博客访客等，简单的定时任务可以直接通过 Spring 中的@Scheduled 注解来实现，复杂的定时任务则可以通过集成 Quartz 来实现，下面分别予以介绍。

13.2.1 @Scheduled

@Scheduled 是由 Spring 提供的定时任务注解，使用方便，配置简单，可以解决工作中大部分的定时任务需求，使用方式如下：

1. 创建工程

首先创建一个普通的 Spring Boot Web 工程，添加 Web 依赖即可，代码如下：

```
1  <dependency>
2      <groupId>org.springframework.boot</groupId>
3      <artifactId>spring-boot-starter-web</artifactId>
4  </dependency>
```

2. 开启定时任务

在项目启动类上添加@EnableScheduling 注解开启定时任务，代码如下：

```
1  @SpringBootApplication
2  @EnableScheduling
3  public class ScheduledApplication {
4      public static void main(String[] args) {
5          SpringApplication.run(ScheduledApplication.class, args);
6      }
7  }
```

3. 配置定时任务

定时任务主要通过@Scheduled注解进行配置，代码如下：

```java
@Component
public class MySchedule {
    @Scheduled(fixedDelay = 1000)
    public void fixedDelay() {
        System.out.println("fixedDelay:"+new Date());
    }
    @Scheduled(fixedRate = 2000)
    public void fixedRate() {
        System.out.println("fixedRate:"+new Date());
    }
    @Scheduled(initialDelay = 1000,fixedRate = 2000)
    public void initialDelay() {
        System.out.println("initialDelay:"+new Date());
    }
    @Scheduled(cron = "0 * * * * ?")
    public void cron() {
        System.out.println("cron:"+new Date());
    }
}
```

代码解释：

- 通过@Scheduled注解来标注一个定时任务，其中 fixedDelay = 1000 表示在当前任务执行结束 1 秒后开启另一个任务，fixedRate = 2000 表示在当前任务开始执行 2 秒后开启另一个定时任务，initialDelay = 1000 则表示首次执行的延迟时间。
- 在@Scheduled注解中也可以使用 cron 表达式，cron = "0 * * * * ?" 表示该定时任务每分钟执行一次。

配置完成后，接下来启动 Spring Boot 项目即可，定时任务部分打印日志如图 13-11 所示。

```
fixedRate:Mon Aug 06 10:20:54 CST 2018
fixedDelay:Mon Aug 06 10:20:54 CST 2018
2018-08-06 10:20:54.878  INFO 3416 --- [
2018-08-06 10:20:54.884  INFO 3416 --- [
initialDelay:Mon Aug 06 10:20:55 CST 2018
fixedDelay:Mon Aug 06 10:20:55 CST 2018
```

图 13-11

13.2.2 Quartz

1. Quartz 简介

Quartz 是一个功能丰富的开源作业调度库，它由 Java 写成，可以集成在任何 Java 应用程序中，包括 Java SE 和 Java EE 等。使用 Quartz 可以创建简单或者复杂的执行计划，它支持数据库、集群、插件以及邮件，并且支持 cron 表达式，具有极高的灵活性。Spring Boot 中集成 Quartz 和 Spring 中集成 Quartz 比较类似，主要提供三个 Bean：JobDetail、Trigger 以及 SchedulerFactory。

2. 整合 Spring Boot

首先创建 Spring Boot 项目，添加 Quartz 依赖，代码如下：

```
1  <dependency>
2      <groupId>org.springframework.boot</groupId>
3      <artifactId>spring-boot-starter-quartz</artifactId>
4  </dependency>
```

然后创建两个 Job，代码如下：

```
1   @Component
2   public class MyFirstJob {
3       public void sayHello() {
4           System.out.println("MyFirstJob:sayHello:"+new Date());
5       }
6   }
7   public class MySecondJob extends QuartzJobBean {
8       private String name;
9       public void setName(String name) {
10          this.name = name;
11      }
12      @Override
13      protected void executeInternal(JobExecutionContext context){
14          System.out.println("hello:"+name+":"+new Date());
15      }
16  }
```

Job 可以是一个普通的 JavaBean，如果是普通的 JavaBean，那么可以先添加 @Component 注解将之注册到 Spring 容器中。Job 也可以继承抽象类 QuartzJobBean，若继承自 QuartzJobBean，则需要实现该类中的 executeInternal 方法，该方法在任务被调用时使用。接下来创建 QuartzConfig 对 JobDetail 和 Trigger 进行配置，代码如下：

```
1   @Configuration
2   public class QuartzConfig {
3       @Bean
4       MethodInvokingJobDetailFactoryBean jobDetail1() {
5           MethodInvokingJobDetailFactoryBean bean =
6               new MethodInvokingJobDetailFactoryBean();
7           bean.setTargetBeanName("myFirstJob");
8           bean.setTargetMethod("sayHello");
9           return bean;
10      }
11      @Bean
12      JobDetailFactoryBean jobDetail2() {
13          JobDetailFactoryBean bean = new JobDetailFactoryBean();
14          bean.setJobClass(MySecondJob.class);
15          JobDataMap jobDataMap = new JobDataMap();
```

```
16              jobDataMap.put("name","sang");
17              bean.setJobDataMap(jobDataMap);
18      bean.setDurability(true);
19              return bean;
20          }
21          @Bean
22          SimpleTriggerFactoryBean simpleTrigger() {
23              SimpleTriggerFactoryBean bean =
24                      new SimpleTriggerFactoryBean();
25              bean.setJobDetail(jobDetail1().getObject());
26              bean.setRepeatCount(3);
27              bean.setStartDelay(1000);
28              bean.setRepeatInterval(2000);
29              return bean;
30          }
31          @Bean
32          CronTriggerFactoryBean cronTrigger() {
33              CronTriggerFactoryBean bean =
34                      new CronTriggerFactoryBean();
35              bean.setJobDetail(jobDetail2().getObject());
36              bean.setCronExpression("* * * * * ?");
37              return bean;
38          }
39          @Bean
40          SchedulerFactoryBean schedulerFactory() {
41              SchedulerFactoryBean bean = new SchedulerFactoryBean();
42              SimpleTrigger simpleTrigger = simpleTrigger().getObject();
43              CronTrigger cronTrigger = cronTrigger().getObject();
44              bean.setTriggers(simpleTrigger,cronTrigger);
45              return bean;
46          }
47      }
```

代码解释：

- JobDetail 的配置有两种方式：第一种方式通过 MethodInvokingJobDetailFactoryBean 类配置 JobDetail，只需要指定 Job 的实例名和要调用的方法即可，注册这种方式无法在创建 JobDetail 时传递参数；第二种方式是通过 JobDetailFactoryBean 来实现的，这种方式只需要指定 JobClass 即可，然后可以通过 JobDataMap 传递参数到 Job 中，Job 中只需要提供属性名，并且提供一个相应的 set 方法即可接收到参数。
- Trigger 有多种不同实现，这里展示两种常用的 Trigger：SimpleTrigger 和 CronTrigger，这两种 Trigger 分别使用 SimpleTriggerFactoryBean 和 CronTriggerFactoryBean 进行创建。
  在 SimpleTriggerFactoryBean 对象中，首先设置 JobDetail，然后通过 setRepeatCount 配置任务循环次数，setStartDelay 配置任务启动延迟时间，setRepeatInterval 配置任务的时间间隔。
  在 CronTriggerFactoryBean 对象中，则主要配置 JobDetail 和 Cron 表达式。
- 最后通过 SchedulerFactoryBean 创建 SchedulerFactory，然后配置 Trigger 即可。

经过这几步的配置，定时任务就配置成功了。接下来启动 Spring Boot 项目，控制台打印日志如图 13-12 所示。

```
hello:sang:Mon Aug 06 23:08:48 CST 2018
2018-08-06 23:08:48.906  INFO 5680 — [
2018-08-06 23:08:48.910  INFO 5680 — [
hello:sang:Mon Aug 06 23:08:49 CST 2018
MyFirstJob:sayHello:Mon Aug 06 23:08:49 CST 2018
hello:sang:Mon Aug 06 23:08:50 CST 2018
hello:sang:Mon Aug 06 23:08:51 CST 2018
MyFirstJob:sayHello:Mon Aug 06 23:08:51 CST 2018
hello:sang:Mon Aug 06 23:08:52 CST 2018
hello:sang:Mon Aug 06 23:08:53 CST 2018
MyFirstJob:sayHello:Mon Aug 06 23:08:53 CST 2018
hello:sang:Mon Aug 06 23:08:54 CST 2018
hello:sang:Mon Aug 06 23:08:55 CST 2018
MyFirstJob:sayHello:Mon Aug 06 23:08:55 CST 2018
hello:sang:Mon Aug 06 23:08:56 CST 2018
hello:sang:Mon Aug 06 23:08:57 CST 2018
hello:sang:Mon Aug 06 23:08:58 CST 2018
hello:sang:Mon Aug 06 23:08:59 CST 2018
```

图 13-12

MyFirstJob 在重复了 3 次之后便不再执行，MySecondJob 则每秒执行一次，一直执行下去。

13.3　批　处　理

13.3.1　Spring Batch 简介

Spring Batch 是一个开源的、全面的、轻量级的批处理框架，通过 Spring Batch 可以实现强大的批处理应用程序的开发。Spring Batch 还提供记录/跟踪、事务管理、作业处理统计、作业重启以及资源管理等功能。Spring Batch 结合定时任务可以发挥更大的作用。

Spring Batch 提供了 ItemReader、ItemProcessor 和 ItemWriter 来完成数据的读取、处理以及写出操作，并且可以将批处理的执行状态持久化到数据库中。接下来通过一个简单的数据复制向读者展示 Spring Boot 中如何使用 Spring Batch。

13.3.2　整合 Spring Boot

现在有一个 data.csv 文件，文件中保存了 4 条用户数据，通过批处理框架读取 data.csv，将之插入数据表中。

首先创建一个 Spring Boot Web 工程，并添加 spring-boot-starter-batch 依赖以及数据库相关依赖，代码如下：

```xml
<dependency>
    <groupId>org.springframework.boot</groupId>
    <artifactId>spring-boot-starter-batch</artifactId>
</dependency>
<dependency>
    <groupId>org.springframework.boot</groupId>
    <artifactId>spring-boot-starter-jdbc</artifactId>
</dependency>
<dependency>
    <groupId>org.springframework.boot</groupId>
    <artifactId>spring-boot-starter-web</artifactId>
</dependency>
<dependency>
    <groupId>com.alibaba</groupId>
    <artifactId>druid</artifactId>
    <version>1.1.10</version>
</dependency>
<dependency>
    <groupId>mysql</groupId>
    <artifactId>mysql-connector-java</artifactId>
</dependency>
```

如前文所说，添加数据库相关依赖是为了将批处理的执行状态持久化到数据库中。项目创建完成后，在 application.properties 中进行数据库基本信息配置：

```
spring.datasource.type=com.alibaba.druid.pool.DruidDataSource
spring.datasource.username=root
spring.datasource.password=root
spring.datasource.url=jdbc:mysql:///batch
spring.datasource.schema=classpath:/org/springframework/batch/core/schema-mysql.sql
spring.batch.initialize-schema=always
spring.batch.job.enabled=false
```

前 4 行配置是数据的基本配置，这里不再赘述。第 5 行配置是项目启动时创建数据表的 SQL 脚本，该脚本由 Spring Batch 提供。第 6 行配置表示在项目启动时执行建表 SQL。第 7 行配置表示禁止 Spring Batch 自动执行。在 Spring Boot 中，默认情况下，当项目启动时就会执行配置好的批处理操作，添加了第 7 行的配置后则不会自动执行，而需要用户手动触发执行，例如发送一个请求，在 Controller 的接口中触发批处理的执行。

接下来在项目启动类上添加@EnableBatchProcessing 注解开启 Spring Batch 支持，代码如下：

```
@SpringBootApplication
@EnableBatchProcessing
public class BatchApplication {
    public static void main(String[] args) {
        SpringApplication.run(BatchApplication.class, args);
    }
}
```

接下来配置批处理，代码如下：

```java
@Configuration
public class CsvBatchJobConfig {
    @Autowired
    JobBuilderFactory jobBuilderFactory;
    @Autowired
    StepBuilderFactory stepBuilderFactory;
    @Autowired
    DataSource dataSource;
    @Bean
    @StepScope
    FlatFileItemReader<User> itemReader() {
        FlatFileItemReader<User> reader = new FlatFileItemReader<>();
        reader.setLinesToSkip(1);
        reader.setResource(new ClassPathResource("data.csv"));
        reader.setLineMapper(new DefaultLineMapper<User>(){{
            setLineTokenizer(new DelimitedLineTokenizer(){{
                setNames("id","username","address","gender");
                setDelimiter("\t");
            }});
            setFieldSetMapper(new BeanWrapperFieldSetMapper<User>(){{
                setTargetType(User.class);
            }});
        }});
        return reader;
    }
    @Bean
    JdbcBatchItemWriter jdbcBatchItemWriter() {
        JdbcBatchItemWriter writer = new JdbcBatchItemWriter();
        writer.setDataSource(dataSource);
        writer.setSql("insert into user(id,username,address,gender) " +
"values(:id,:username,:address,:gender)");
        writer.setItemSqlParameterSourceProvider(
            new BeanPropertyItemSqlParameterSourceProvider<>());
        return writer;
    }
    @Bean
    Step csvStep() {
        return stepBuilderFactory.get("csvStep")
            .<User, User>chunk(2)
            .reader(itemReader())
            .writer(jdbcBatchItemWriter())
            .build();
    }
    @Bean
    Job csvJob() {
        return jobBuilderFactory.get("csvJob")
            .start(csvStep())
            .build();
    }
}
```

代码解释：

- 创建 CsvBatchJobConfig 进行 Spring Batch 配置，同时注入 JobBuilderFactory、StepBuilderFactory 以及 DataSource 备用，其中 JobBuilderFactory 将用来构建 Job，StepBuilderFactory 用来构建 Step，DataSource 则用来支持持久化操作，这里持久化方案是 Spring-Jdbc。
- 第 9~25 行配置一个 ItemReader，Spring Batch 提供了一些常用的 ItemReader，例如 JdbcPagingItemReader 用来读取数据库中的数据，StaxEventItemReader 用来读取 XML 数据，本案例中的 FlatFileItemReader 则是一个加载普通文件的 ItemReader。在 FlatFileItemReader 的配置过程中，由于 data.csv 文件第一行是标题，因此通过 setLinesToSkip 方法设置跳过一行，然后通过 setResource 方法配置 data.csv 文件的位置，笔者的 data.csv 文件放在 classpath 目录下，然后通过 setLineMapper 方法设置每一行的数据信息，setNames 方法配置了 data.csv 文件一共有 4 列，分别是 id、username、address 以及 gender，setDelimiter 则是配置列与列之间的间隔符（将通过间隔符对每一行的数据进行切分），最后设置要映射的实体类属性即可。
- 第 26~35 行配置 ItemWriter，即数据的写出逻辑，Spring Batch 也提供了多个 ItemWriter 的实现，常见的如 FlatFileItemWriter，表示将数据写出为一个普通文件，StaxEventItemWriter 表示将数据写出为 XML。另外，还有针对不同数据库提供的写出操作支持类，如 MongoItemWriter、JpaItemWriter、Neo4jItemWriter 以及 HibernateItemWriter 等，本案例使用的 JdbcBatchItemWriter 则是通过 JDBC 将数据写出到一个关系型数据库中。JdbcBatchItemWriter 主要配置数据以及数据插入 SQL，注意占位符的写法是"：属性名"。最后通过 BeanPropertyItemSqlParameterSourceProvider 实例将实体类的属性和 SQL 中的占位符一一映射。
- 第 36~43 行配置一个 Step，Step 通过 stepBuilderFactory 进行配置，首先通过 get 获取一个 StepBuilder，get 方法的参数就是该 Step 的 name，然后调用 chunk 方法的参数 2，表示每读取到两条数据就执行一次 write 操作，最后分别配置 reader 和 writer。
- 第 44~49 行配置一个 Job，通过 jobBuilderFactory 构建一个 Job，get 方法的参数为 Job 的 name，然后配置该 Job 的 Step 即可。

当然，这里还涉及一个 User 实体类，代码如下：

```
public class User {
    private Integer id;
    private String username;
    private String address;
    private String gender;
    //省略getter/setter
}
```

另外，classpath 下的 data.csv 文件如图 13-13 所示。

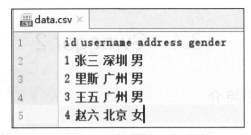

图 13-13

接下来创建 Controller，当用户发起一个请求时触发批处理，代码如下：

```
@RestController
public class HelloController {
    @Autowired
    JobLauncher jobLauncher;
    @Autowired
    Job job;
    @GetMapping("/hello")
    public void hello() {
        try {
            jobLauncher.run(job, new JobParametersBuilder().toJobParameters());
        } catch (Exception e) {
            e.printStackTrace();
        }
    }
}
```

JobLauncher 由框架提供，Job 则是刚刚配置的，通过调用 JobLauncher 中的 run 方法启动一个批处理。

最后根据上文的实体类在数据库中创建一个 user 表，然后启动 Spring Boot 工程并访问 http://localhost:8080/hello 接口，访问成功后，batch 库中会自动创建出多个批处理相关的表，如图 13-14 所示。这些表用来记录批处理的执行状态，同时，data.csv 中的数据也已经成功插入 user 表中，如图 13-15 所示。

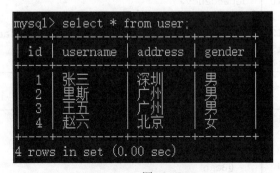

图 13-14　　　　　　　　　　　　　图 13-15

13.4　Swagger 2

13.4.1　Swagger 2 简介

在前后端分离开发中，为了减少与其他团队的沟通成本，一般构建一份 RESTful API 文档来描述所有的接口信息，但是这种做法有很大的弊端，分别说明如下：

- 接口众多，编写 RESTful API 文档工作量巨大，因为 RESTful API 文档不仅要包含接口的基本信息，如接口地址、接口请求参数以及接口返回值等，还要包含 HTTP 请求类型、HTTP 请求头、请求参数类型、返回值类型、所需权限等。
- 维护不方便，一旦接口发生变化，就要修改文档。
- 接口测试不方便，一般只能借助第三方工具（如 Postman）来测试。

Swagger 2 是一个开源软件框架，可以帮助开发人员设计、构建、记录和使用 RESTful Web 服务，它将代码和文档融为一体，可以完美解决上面描述的问题，使开发人员将大部分精力集中到业务中，而不是繁杂琐碎的文档中。

Swagger 2 可以非常轻松地整合到 Spring Boot 项目中，下面来看如何整合。

13.4.2　整合 Spring Boot

首先创建 Spring Boot Web 项目，添加 Swagger 2 相关依赖，代码如下：

```
1   <dependency>
2       <groupId>org.springframework.boot</groupId>
3       <artifactId>spring-boot-starter-web</artifactId>
4   </dependency>
5   <dependency>
6       <groupId>io.springfox</groupId>
7       <artifactId>springfox-swagger2</artifactId>
8       <version>2.9.2</version>
9   </dependency>
10  <dependency>
11      <groupId>io.springfox</groupId>
12      <artifactId>springfox-swagger-ui</artifactId>
13      <version>2.9.2</version>
14  </dependency>
```

接下来创建 Swagger 2 的配置类，代码如下：

```
1   @Configuration
2   @EnableSwagger2
3   public class SwaggerConfig {
4       @Bean
5       Docket docket() {
```

```
 6              return new Docket(DocumentationType.SWAGGER_2)
 7                      .select()
 8                      .apis(RequestHandlerSelectors.basePackage("org.sang.controller"))
 9                      .paths(PathSelectors.any())
10                      .build().apiInfo(new ApiInfoBuilder()
11                      .description("微人事接口测试文档")
12                      .contact(new Contact("江南一点雨",
13  "https://github.com/lenve",
14  "wangsong0210@gmail.com"))
15                      .version("v1.0")
16                      .title("API 测试文档")
17                      .license("Apache2.0")
18                      .licenseUrl("http://www.apache.org/licenses/LICENSE-2.0")
19                      .build());
20      }
21  }
```

代码解释：

- 首先通过@EnableSwagger2 注解开启 Swagger 2，然后最主要的是配置一个 Docket。
- 通过 apis 方法配置要扫描的 controller 位置，通过 paths 方法配置路径。
- 在 apiInfo 中构建文档的基本信息，例如描述、联系人信息、版本、标题等。

Swagger 2 配置完成后，接下来就可以开发接口了，代码如下：

```
 1  @RestController
 2  @Api(tags = "用户数据接口")
 3  public class UserController {
 4      @ApiOperation(value = "查询用户", notes = "根据 id 查询用户")
 5      @ApiImplicitParam(paramType = "path",name = "id",
 6              value = "用户id",required = true)
 7      @GetMapping("/user/{id}")
 8      public String getUserById(@PathVariable Integer id) {
 9          return "/user/" + id;
10      }
11      @ApiResponses({
12      @ApiResponse(code = 200,message = "删除成功!"),
13      @ApiResponse(code = 500,message = "删除失败!")})
14      @ApiOperation(value = "删除用户", notes = "通过 id 删除用户")
15      @DeleteMapping("/user/{id}")
16      public Integer deleteUserById(@PathVariable Integer id) {
17          return id;
18      }
19      @ApiOperation(value = "添加用户",
20              notes = "添加一个用户，传入用户名和地址")
21      @ApiImplicitParams({
22      @ApiImplicitParam(paramType = "query", name = "username",
23              value = "用户名", required = true,defaultValue = "sang"),
24      @ApiImplicitParam(paramType = "query", name = "address",
25              value = "用户地址", required = true,defaultValue = "shenzhen")})
26      @PostMapping("/user")
```

```
27        public String addUser(@RequestParam String username,
28                              @RequestParam String address) {
29            return username + ":" + address;
30        }
31        @ApiOperation(value = "修改用户", notes = "修改用户,传入用户信息")
32        @PutMapping("/user")
33        public String updateUser(@RequestBody User user) {
34            return user.toString();
35        }
36  @GetMapping("/ignore")
37  @ApiIgnore
38  public void ingoreMethod() {
39  }
40  }
41  @ApiModel(value = "用户实体类",description = "用户信息描述类")
42  public class User {
43      @ApiModelProperty(value = "用户名")
44      private String username;
45      @ApiModelProperty(value = "用户地址")
46      private String address;
47      //省略getter/setter
48  }
```

代码解释：

- @Api 注解用在类上，用来描述整个 Controller 信息。
- @ApiOperation 注解用在开发方法上，用来描述一个方法的基本信息，value 是对方法作用的一个简短描述，notes 则用来备注该方法的详细作用。
- @ApiImplicitParam 注解用在方法上，用来描述方法的参数，paramType 是指方法参数的类型，可选值有 path（参数获取方式@PathVariable）、query（参数获取方式@RequestParam）、header（参数获取方式@RequestHeader）、body 以及 form；name 表示参数名称，和参数变量对应；value 是参数的描述信息；required 表示该字段是否必填；defaultValue 表示该字段的默认值。注意，这里的 required 和 defaultValue 等字段只是文档上的约束描述，并不能真正约束接口，约束接口还需要在@RequestParam 中添加相关属性。
  如果方法有多个参数，可以将多个参数的@ApiImplicitParam 注解放到@ApiImplicitParams 中。
- @ApiResponse 注解是对响应结果的描述，code 表示响应码，message 表示相应的描述信息，若有多个@ApiResponse，则放在一个@ApiResponses 中。
- 在 updateUser 方法中，使用@RequestBody 注解来接收数据，此时可以通过@ApiModel 注解和@ApiModelProperty 注解配置 User 对象的描述信息。
- @ApiIgnore 注解表示不对某个接口生成文档。

接下来启动 Spring Boot 项目，在浏览器中输入 http://localhost:8080/swagger-ui.html 即可看到接口文档，如图 13-16 所示。

图 13-16

展开用户数据接口，即可看到所有接口的描述，如图 13-17 所示。

展开一个接口描述，内容如图 13-18 所示，单击 Try it out 按钮，可以实现对该接口的测试。

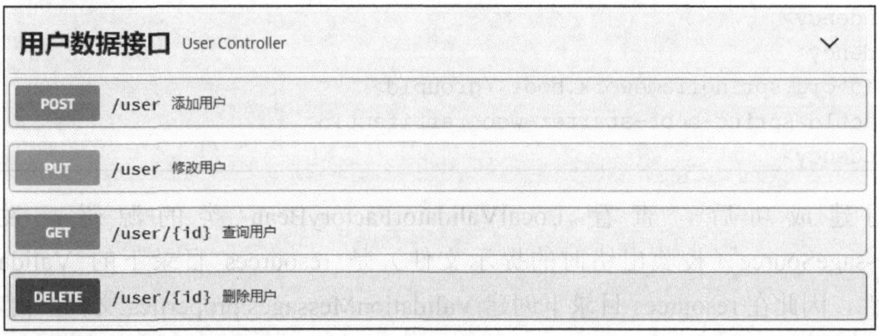

图 13-17

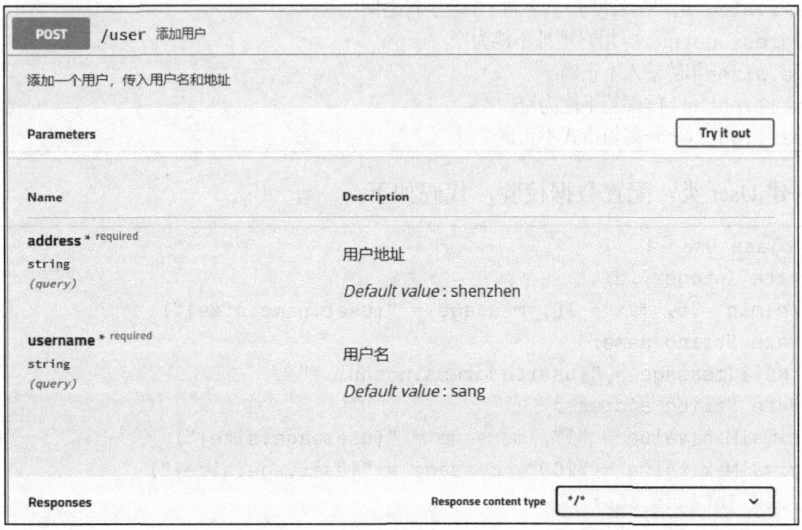

图 13-18

13.5 数据校验

数据校验是开发过程中一个常见的环节，一般来说，为了提高系统运行效率，都会在前端进行数据校验，但是这并不意味着不必在后端做数据校验了，因为用户还是可能在获取数据接口后手动传入非法数据，所以后端还是需要做数据校验。Spring Boot 对此也提供了相关的自动化配置解决方案，下面分别予以介绍。

13.5.1 普通校验

普通校验是基础用法，非常容易，首先需要用户在 Spring Boot Web 项目中添加数据校验相关的依赖，代码如下：

```xml
<dependency>
<groupId>org.springframework.boot</groupId>
<artifactId>spring-boot-starter-validation</artifactId>
</dependency>
<dependency>
<groupId>org.springframework.boot</groupId>
<artifactId>spring-boot-starter-web</artifactId>
</dependency>
```

项目创建成功后，查看 LocalValidatorFactoryBean 类的源码，发现默认的 ValidationMessageSource（校验出错时的提示文件）是 resources 目录下的 ValidationMessages.properties 文件，因此在 resources 目录下创建 ValidationMessages.properties 文件，内容如下（如果文件出现乱码，参考 2.6 节解决）：

```
user.name.size=用户名长度介于5到10个字符之间
user.address.notnull=用户地址不能为空
user.age.size=年龄输入不正确
user.email.notnull=邮箱不能为空
user.email.pattern=邮箱格式不正确
```

接下来创建 User 类，配置数据校验，代码如下：

```java
public class User {
    private Integer id;
    @Size(min = 5, max = 10, message = "{user.name.size}")
    private String name;
    @NotNull(message = "{user.address.notnull}")
    private String address;
    @DecimalMin(value = "1", message = "{user.age.size}")
    @DecimalMax(value = "200", message = "{user.age.size}")
    private Integer age;
    @Email(message = "{user.email.pattern}")
    @NotNull(message = "{user.email.notnull}")
```

```
12      private String email;
13      //省略 getter/setter
14  }
```

代码解释：

- @Size 表示一个字符串的长度或者一个集合的大小，必须在某一个范围中；min 参数表示范围的下限；max 参数表示范围的上限；message 表示校验失败时的提示信息。
- @NotNull 注解表示该字段不能为空。
- @DecimalMin 注解表示对应属性值的下限，@DecimalMax 注解表示对应属性值的上限。
- @Email 注解表示对应属性格式是一个 Email。

配置完成后，接下来创建 UserController，代码如下：

```
1   @RestController
2   public class UserController {
3       @PostMapping("/user")
4       public List<String> addUser(@Validated User user, BindingResult result) {
5           List<String> errors = new ArrayList<>();
6           if (result.hasErrors()) {
7               List<ObjectError> allErrors = result.getAllErrors();
8               for (ObjectError error : allErrors) {
9                   errors.add(error.getDefaultMessage());
10              }
11          }
12          return errors;
13      }
14  }
```

代码解释：

- 给 User 参数添加@Validated 注解，表示需要对该参数做校验，紧接着的 BindingResult 参数表示在校验出错时保存的出错信息。
- 如果 BindingResult 中的 hasErrors 方法返回 true，表示有错误信息，此时遍历错误信息，将之返回给前端。

配置完成后，接下来使用 Postman 进行测试，例如直接访问 "/user" 接口，结果如图 13-19 所示。

图 13-19

如果传入用户地址、一个非法邮箱地址以及一个格式不正确的用户名，结果如图 13-20 所示。

图 13-20

13.5.2 分组校验

有的时候,开发者在某一个实体类中定义了很多校验规则,但是在某一次业务处理中,并不需要这么多校验规则,此时就可以使用分组校验。分组校验步骤如下:

首先创建两个分组接口,代码如下:

```
public interface ValidationGroup1 {
}
public interface ValidationGroup2 {
}
```

然后在实体类中添加分组信息,代码如下:

```
public class User {
    private Integer id;
    @Size(min = 5, max = 10, message = "{user.name.size}",
            groups = ValidationGroup1.class)
    private String name;
    @NotNull(message = "{user.address.notnull}",groups = ValidationGroup2.class)
    private String address;
    @DecimalMin(value = "1", message = "{user.age.size}")
    @DecimalMax(value = "200", message = "{user.age.size}")
    private Integer age;
    @Email(message = "{user.email.pattern}")
    @NotNull(message = "{user.email.notnull}",
            groups = {ValidationGroup1.class,ValidationGroup2.class})
    private String email;
    //省略getter/setter
}
```

这次在部分注解中添加了 groups 属性,表示该校验规则所属的分组,接下来在@Validated 注解中指定校验分组,代码如下:

```
@RestController
public class UserController {
    @PostMapping("/user")
    public List<String> addUser(@Validated(ValidationGroup2.class) User user,
                        BindingResult result) {
        List<String> errors = new ArrayList<>();
```

```
7              if (result.hasErrors()) {
8                  List<ObjectError> allErrors = result.getAllErrors();
9                  for (ObjectError error : allErrors) {
10                     errors.add(error.getDefaultMessage());
11                 }
12             }
13             return errors;
14         }
15 }
```

@Validated(ValidationGroup2.class)表示这里的校验使用 ValidationGroup2 分组的校验规则，即只校验邮箱地址是否为空、用户地址是否为空。测试案例如图 13-21 所示。在图 13-21 的测试案例中，虽然邮箱地址格式不正确，name 长度也不满足要求，但是这些都不属于 ValidationGroup2 校验分组的校验规则，这里只校验邮箱地址是否为空、用户地址是否为空。

图 13-21

13.5.3 校验注解

前面向读者介绍了几个常见的校验注解，实际上校验注解不止前面提到的几个，完整的校验注解可参考表 13-1。

表 13-1 校验注解

校验注解	注解的元素类型	描述
AssertFalse	Boolean、boolean	被注解的元素值必须为 false
AssertTrue	Boolean、boolean	被注解的元素值必须为 true
DecimalMax	BigDecimal、BigInteger、CharSequence、byte、short、int、long 以及它们各自的包装类	被注解的元素值小于等于@DecimalMax 注解中的 value 值
DecimalMin	BigDecimal、BigInteger、CharSequence、byte、short、int、long 以及它们各自的包装类	被注解的元素值大于等于@DecimalMin 注解中的 value 值
Max	BigDecimal、BigInteger、byte、short、int、long 以及它们各自的包装类	被注解的元素值小于等于@ Max 注解中的 value 值
Min	BigDecimal、BigInteger、byte、short、int、long 以及它们各自的包装类	被注解的元素值大于等于@ Min 注解中的 value 值
Digits	BigDecimal、BigInteger、CharSequence、byte、short、int、long 以及它们各自的包装类	被注解的元素必须是一个数字，其值必须在可接受的范围内（整数位数和小数位数在指定范围内）

（续表）

校验注解	注解的元素类型	描述
Email	CharSequence	被注解的元素值必须是 Email 格式
Future	java.util.Date、java.util.Calendar 以及 java.time 包下的时间类	被注解的元素值必须是一个未来的日期
Past	java.util.Date、java.util.Calendar 以及 java.time 包下的时间类	被注解的元素值必须是一个过去的日期
PastOrPresent	java.util.Date、java.util.Calendar 以及 java.time 包下的时间类	被注解的元素值必须是一个过去的日期或者当前日期
FutureOrPresent	java.util.Date、java.util.Calendar 以及 java.time 包下的时间类	被注解的元素值必须是一个未来的日期或当前日期
Negative	BigDecimal、BigInteger、byte、short、int、long 以及它们各自的包装类	被注解的元素必须是负数
NegativeOrZero	BigDecimal、BigInteger、byte、short、int、long 以及它们各自的包装类	被注解的元素必须是负数或 0
Positive	BigDecimal、BigInteger、byte、short、int、long 以及它们各自的包装类	被注解的元素必须是正数
PositiveOrZero	BigDecimal、BigInteger、byte、short、int、long 以及它们各自的包装类	被注解的元素必须是正数或 0
NotBlank	CharSequence	被注解的元素必须不为 null 并且至少有一个非空白的字符
NotEmpty	CharSequence、Collection、Map、Array	被注解的字符串不为 null 或空字符串，被注解的集合或数组不为空。和 @NotBlank 注解相比，一个空格字符串在 @NotBlank 验证不通过，但是在 @NotEmpty 中验证通过
NotNull	任意类型	被注解的元素不为 null
Null	任意类型	被注解的元素为 null
Pattern	CharSequence	被注解的元素必须符合指定的正则表达式
Size	CharSequence、Collection、Map、Array	被注解的字符串长度、集合或者数组的大小必须在指定范围内

13.6 小　　结

本章向读者介绍了企业开发中一些常用的功能，如邮件发送、定时任务、批处理、Swagger 2 以及数据校验，这些功能都有非常广泛的使用场景，如用户注册、修改密码、定时备份、接口文档等，除了 Swagger 2 外，其他 4 个功能在 Spring Boot 中都提供了相关的 Starter，简化了开发者的使用步骤，提高了开发效率。

第 14 章

应用监控

本章概要

- 监控端点配置
- 监控信息可视化
- 邮件报警

当一个 Spring Boot 项目运行时,开发者需要对 Spring Boot 项目进行实时监控,获取项目的运行情况,在项目出错时能够实现自动报警等。Spring Boot 提供了 actuator 来帮助开发者获取应用程序的实时运行数据。开发者可以选择使用 HTTP 端点或 JMX 来管理和监控应用程序,获取应用程序的运行数据,包括健康状况、应用信息、内存使用情况等。

14.1 端点配置

14.1.1 开启端点

在 Spring Boot 中开启应用监控非常容易,只需要添加 spring-boot-starter-actuator 依赖即可,actuator(执行器)是制造业术语,指一个用于移动或控制机械装置的工具,一个很小的变化就能让执行器产生大量的运动。依赖如下:

```
1  <dependency>
2      <groupId>org.springframework.boot</groupId>
3      <artifactId>spring-boot-starter-actuator</artifactId>
```

```
4        </dependency>
```

开发者可以使用执行器中的端点（EndPoints）对应用进行监控或者与应用进行交互，Spring Boot 默认包含许多端点，如表 14-1 所示。

表 14-1　Spring Boot 默认包含的端点

端点	端点描述	是否开启
auditevents	展示当前应用程序的审计事件信息	Yes
beans	展示所有 Spring Beans 信息	Yes
conditions	展示一个自动配置类的使用报告，该报告展示所有自动配置类及它们被使用或未被使用的原因	Yes
configprops	展示所有 @ConfigurationProperties 的列表	Yes
env	展示系统运行环境信息	Yes
flyway	展示数据库迁移路径	Yes
health	展示应用程序的健康信息	Yes
httptrace	展示 trace 信息（默认为最新的 100 条 HTTP 请求）	Yes
info	展示应用的定制信息，这些定制信息以 info 开头	Yes
loggers	展示并修改应用的日志配置	Yes
liquibase	展示任何 Liquibase 数据库迁移路径	Yes
metrics	展示应用程序度量信息	Yes
mappings	展示所有 @RequestMapping 路径的集合列表	Yes
scheduledtasks	展示应用的所有定时任务	Yes
shutdown	远程关闭应用接口	No
sessions	展示并操作 Spring Session 会话	Yes
threaddump	展示线程活动的快照	Yes

如果是一个 Web 应用，还会有表 14-2 所示的端点。

表 14-2　Web 应用另外包含的端点

端点	端点描述	是否开启
heapdump	返回一个 GZip 压缩的 hprof 堆转储文件	Yes
jolokia	展示通过 HTTP 暴露的 JMX beans	Yes
logfile	返回日志文件内容	Yes
prometheus	展示一个可以被 Prometheus 服务器抓取的 metrics 数据	Yes

这些端点大部分都是默认开启的，只有 shutdown 端点默认未开启，如果需要开启，可以在 application.properties 中通过如下配置开启：

```
1   management.endpoint.shutdown.enabled=true
```

如果开发者不想暴露这么多端点，那么可以关闭默认的配置，然后手动指定需要开启哪些端点，如下配置表示关闭所有端点，只开启 info 端点：

```
1   management.endpoints.enabled-by-default=false
2   management.endpoint.info.enabled=true
```

14.1.2 暴露端点

由于有的端点包含敏感信息，因此端点启用和暴露是两码事。表 14-3 展示了端点的默认暴露情况。

表 14-3 端点的默认暴露情况

端点	JMX	Web
auditevents	Yes	No
beans	Yes	No
conditions	Yes	No
configprops	Yes	No
env	Yes	No
flyway	Yes	No
health	Yes	Yes
httptrace	Yes	No
info	Yes	Yes
loggers	Yes	No
liquibase	Yes	No
metrics	Yes	No
mappings	Yes	No
scheduledtasks	Yes	No
shutdown	Yes	No
sessions	Yes	No
threaddump	Yes	No
heapdump	N/A	No
jolokia	N/A	No
logfile	N/A	No
prometheus	N/A	No

在 Web 应用中，默认只有 health 和 info 两个端点暴露，即当开发者在 Spring Boot 项目中加入 spring-boot-starter-actuator 依赖并启动 Spring Boot 项目后，默认只有这两个端口可访问，启动日志如图 14-1 所示。

```
main] s.b.a.e.w.s.WebMvcEndpointHandlerMapping : Mapped "{[/actuator/health],methods=[GET],produces=
main] s.b.a.e.w.s.WebMvcEndpointHandlerMapping : Mapped "{[/actuator/info],methods=[GET],produces=[a
main] s.b.a.e.w.s.WebMvcEndpointHandlerMapping : Mapped "{[/actuator],methods=[GET],produces=[applic
```

图 14-1

开发者可以在配置文件中自定义需要暴露哪些端点，例如要暴露 mappings 和 metrics 端点，添加如下配置即可：

| 1 | management.endpoints.web.exposure.include=mappings,metrics |

如果要暴露所有端点，添加如下配置即可：

```
1  management.endpoints.web.exposure.include=*
```

由于*在 YAML 格式的配置文件中有特殊的含义，因此如果在 YAML 中配置暴露所有端点，配置方式如下：

```
1  management:
2    endpoints:
3      web:
4        exposure:
5          include: "*"
```

当配置暴露所有端点后，启动日志如图 14-2 所示。

```
s.b.a.e.w.s.WebMvcEndpointHandlerMapping : Mapped "{[/actuator/auditevents],methods=[GET],produc
s.b.a.e.w.s.WebMvcEndpointHandlerMapping : Mapped "{[/actuator/beans],methods=[GET],produces=[ap
s.b.a.e.w.s.WebMvcEndpointHandlerMapping : Mapped "{[/actuator/health],methods=[GET],produces=[a
s.b.a.e.w.s.WebMvcEndpointHandlerMapping : Mapped "{[/actuator/conditions],methods=[GET],produce
s.b.a.e.w.s.WebMvcEndpointHandlerMapping : Mapped "{[/actuator/configprops],methods=[GET],produ
s.b.a.e.w.s.WebMvcEndpointHandlerMapping : Mapped "{[/actuator/env],methods=[GET],produces=[appl
s.b.a.e.w.s.WebMvcEndpointHandlerMapping : Mapped "{[/actuator/env/{toMatch}],methods=[GET],proc
s.b.a.e.w.s.WebMvcEndpointHandlerMapping : Mapped "{[/actuator/info],methods=[GET],produces=[app
s.b.a.e.w.s.WebMvcEndpointHandlerMapping : Mapped "{[/actuator/loggers],methods=[GET],produces=[
s.b.a.e.w.s.WebMvcEndpointHandlerMapping : Mapped "{[/actuator/loggers/{name}],methods=[GET],pro
s.b.a.e.w.s.WebMvcEndpointHandlerMapping : Mapped "{[/actuator/loggers/{name}],methods=[POST],co
s.b.a.e.w.s.WebMvcEndpointHandlerMapping : Mapped "{[/actuator/heapdump],methods=[GET],produces=
s.b.a.e.w.s.WebMvcEndpointHandlerMapping : Mapped "{[/actuator/threaddump],methods=[GET],produce
s.b.a.e.w.s.WebMvcEndpointHandlerMapping : Mapped "{[/actuator/metrics/{requiredMetricName}],met
s.b.a.e.w.s.WebMvcEndpointHandlerMapping : Mapped "{[/actuator/metrics],methods=[GET],produces=[
s.b.a.e.w.s.WebMvcEndpointHandlerMapping : Mapped "{[/actuator/scheduledtasks],methods=[GET],pro
s.b.a.e.w.s.WebMvcEndpointHandlerMapping : Mapped "{[/actuator/httptrace],methods=[GET],produces
s.b.a.e.w.s.WebMvcEndpointHandlerMapping : Mapped "{[/actuator/mappings],methods=[GET],produces=
s.b.a.e.w.s.WebMvcEndpointHandlerMapping : Mapped "{[/actuator],methods=[GET],produces=[applicat
```

图 14-2

此时读者发现，并非所有的端点都在启动日志中展示出来了，这是因为部分端点需要相关依赖才能使用，例如 sessions 端点需要 spring-session 依赖。对于已经展示出来的接口，开发者可以直接发送相应的请求查看相关信息，例如请求 health 端点，如图 14-3 所示。"status":"up"表示应用在线，默认展示的 health 信息较少，后面会详细介绍 health 端点的其他配置。

图 14-3

14.1.3 端点保护

如果这些端点需要对外提供服务，那么最好能够将这些端点保护起来，若 classpath 中存在 Spring Security，则默认使用 Spring Security 保护，使用 Spring Security 保护的步骤很简单，首先添加 Spring Security 依赖：

```
1  <dependency>
2      <groupId>org.springframework.boot</groupId>
3      <artifactId>spring-boot-starter-security</artifactId>
4  </dependency>
```

然后添加 Spring Security 配置，代码如下：

```
1  @Configuration
2  public class ActuatorSecurity extends WebSecurityConfigurerAdapter {
3      @Override
4      protected void configure(HttpSecurity http) throws Exception {
5          http.requestMatcher(EndpointRequest.toAnyEndpoint())
6              .authorizeRequests()
7                  .anyRequest().hasRole("ADMIN")
8                  .and()
9              .httpBasic();
10     }
11 }
```

在 HttpSecurity 中配置所有的 Endpoint 都需要具有 ADMIN 角色才能访问，同时开启 HttpBasic 认证。注意，EndpointRequest.toAnyEndpoint()表示匹配所有的 Endpoint，例如 shutdown、mappings、health 等，但是不包括开发者通过@RequestMapping 注解定义的接口（关于 Spring Security 的更多信息，可以参考第 10 章）。

这里为了演示方便，Spring Security 就不连接数据库了，直接在 application.properties 中定义一个用户进行测试，代码如下：

```
1  spring.security.user.name=sang
2  spring.security.user.password=123
3  spring.security.user.roles=ADMIN
```

定义完成后，启动 Spring Boot 项目，再去访问 health 端点，需要登录后才可以访问。

14.1.4 端点响应缓存

对于一些不带参数的端点请求会自动进行缓存，开发者可通过如下方式配置缓存时间：

```
1  management.endpoint.beans.cache.time-to-live=100s
```

这个配置表示 beans 端点的缓存时间为 100s，如果要配置其他端点，只需将 beans 修改为其他端点名称即可。注意，如果端点添加了 Spring Security 保护，此时 Principal 会被视为端点的输入，因此端点响应将不被缓存。

14.1.5 路径映射

默认情况下，所有端点都暴露在“/actuator”路径下，例如 health 端点的访问路径是“/actuator/health”，如果开发者需要对端点路径进行定制，可通过如下配置进行：

```
1 management.endpoints.web.base-path=/
2 management.endpoints.web.path-mapping.health=healthcheck
```

第一行配置表示将“/actuator”修改为“/”，这行配置会使所有的端点访问路径失去“/actuator”前缀；第二行配置表示将“health”修改为“healthcheck”，修改后 health 端点的访问路径由“/actuator/health”变为“/healthcheck”。此时启动项目，启动日志如图 14-4 所示。

```
s.b.a.e.w.s.WebMvcEndpointHandlerMapping : Mapped "{[/auditevents],methods=[GET],produces=[app
s.b.a.e.w.s.WebMvcEndpointHandlerMapping : Mapped "{[/beans],methods=[GET],produces=[applicat
s.b.a.e.w.s.WebMvcEndpointHandlerMapping : Mapped "{[/healthcheck],methods=[GET],produces=[app
s.b.a.e.w.s.WebMvcEndpointHandlerMapping : Mapped "{[/conditions],methods=[GET],produces=[app
s.b.a.e.w.s.WebMvcEndpointHandlerMapping : Mapped "{[/configprops],methods=[GET],produces=[app
s.b.a.e.w.s.WebMvcEndpointHandlerMapping : Mapped "{[/env],methods=[GET],produces=[applicatio
s.b.a.e.w.s.WebMvcEndpointHandlerMapping : Mapped "{[/env/{toMatch}],methods=[GET],produces=[a
s.b.a.e.w.s.WebMvcEndpointHandlerMapping : Mapped "{[/info],methods=[GET],produces=[applicati
s.b.a.e.w.s.WebMvcEndpointHandlerMapping : Mapped "{[/loggers],methods=[GET],produces=[applica
s.b.a.e.w.s.WebMvcEndpointHandlerMapping : Mapped "{[/loggers/{name}],methods=[GET],produces=
s.b.a.e.w.s.WebMvcEndpointHandlerMapping : Mapped "{[/loggers/{name}],methods=[POST],consumes=
s.b.a.e.w.s.WebMvcEndpointHandlerMapping : Mapped "{[/heapdump],methods=[GET],produces=[appli
s.b.a.e.w.s.WebMvcEndpointHandlerMapping : Mapped "{[/threaddump],methods=[GET],produces=[app
s.b.a.e.w.s.WebMvcEndpointHandlerMapping : Mapped "{[/metrics/{requiredMetricName}],methods=[G
s.b.a.e.w.s.WebMvcEndpointHandlerMapping : Mapped "{[/metrics],methods=[GET],produces=[applica
s.b.a.e.w.s.WebMvcEndpointHandlerMapping : Mapped "{[/scheduledtasks],methods=[GET],produces=
s.b.a.e.w.s.WebMvcEndpointHandlerMapping : Mapped "{[/httptrace],methods=[GET],produces=[appl
s.b.a.e.w.s.WebMvcEndpointHandlerMapping : Mapped "{[/mappings],methods=[GET],produces=[appli
```

图 14-4

14.1.6 CORS 支持

关于 CORS 的介绍，读者可以参考 4.6 节。

所有端点默认都没有开启跨域，开发者可以通过如下配置快速开启 CORS 支持，进而实现跨域：

```
management.endpoints.web.cors.allowed-origins=http://localhost:8081
management.endpoints.web.cors.allowed-methods=GET,POST
```

这个配置表示允许端点处理来自 http://localhost:8081 地址的请求，允许的请求方法为 GET 和 POST。

14.1.7 健康信息

1. 展示健康信息详情

开发者可以通过查看健康信息来获取应用的运行数据,进而提早发现应用问题,提早解决,避免造成损失。默认情况下,开发者只能获取到 status 信息(见图 14-3),这是因为 detail 信息默认不显示,开发者可以通过 management.endpoint.health.show-details 属性来配置 detail 信息的显示策略,该属性的取值一共有三种:

- never 即不显示 details 信息,默认即此。
- when-authorized   details 信息只展示给认证用户,即用户登录后可以查看 details 信息,未登录则不能查看,另外还可以通过 management.endpoint.health.roles 属性配置要求的角色,如果不配置,那么通过认证的用户都可以查看 details 信息,如果配置了,例如 management.endpoint.health.roles=ADMIN 表示认证的用户必须具有 ADMIN 角色才能查看 details 信息。
- always   将 details 信息展示给所有用户。

例如,在 pom.xml 文件中引入 Spring Security 后,在 application.properties 文件中增加如下配置:

```
1  management.endpoints.web.exposure.include=*
2  management.endpoint.health.show-details=when_authorized
3  management.endpoint.health.roles=ADMIN
4  spring.security.user.name=sang
5  spring.security.user.password=123
6  spring.security.user.roles=ADMIN
```

这里首先暴露所有的端点,配置 health 的 details 信息只展示给认证用户,并且认证用户还要具有 ADMIN 角色,然后配置一个默认的用户,用户名是 sang,用户密码是 123,用户角色是 ADMIN。

配置完成后,启动 Spring Boot 项目,在 Postman 中访问 health 端点,如图 14-5 所示。

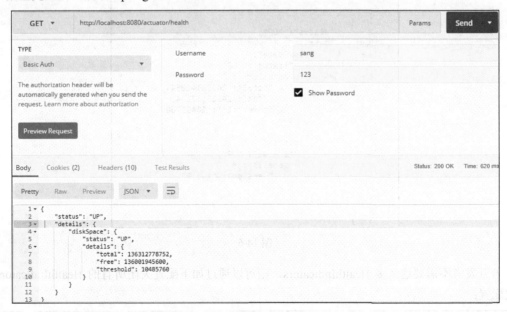

图 14-5

2. 健康指示器

Spring Boot 会根据 classpath 中依赖的添加情况来自动配置一些 HealthIndicators，如表 14-4 所示。

表 14-4 自动配置的 HealthIndicators

名字	描述
CassandraHealthIndicator	检查 Cassandra 数据库状况
DiskSpaceHealthIndicator	低磁盘空间检查
DataSourceHealthIndicator	检查是否可以从 DataSource 获取一个 Connection
ElasticsearchHealthIndicator	检查 Elasticsearch 集群状况
InfluxDbHealthIndicator	检查 InfluxDB 状况
JmsHealthIndicator	检查 JMS 消息代理状况
MailHealthIndicator	检查邮件服务器状况
MongoHealthIndicator	检查 MongoDB 数据库状况
Neo4jHealthIndicator	检查 Neo4j 服务器状况
RabbitHealthIndicator	检查 Rabbit 服务器状况
RedisHealthIndicator	检查 Redis 服务器状况
SolrHealthIndicator	检查 Solr 服务器状况

如果项目中存在相关的依赖，那么列表中对应的 HealthIndicators 将会被自动配置，例如在 pom.xml 文件中添加了 Redis 依赖，此时访问 health 端点，结果如图 14-6 所示。

图 14-6

若开发者不需要这么多 HealthIndicators，则可以通过如下配置关闭所有的 HealthIndicators 自动化配置：

```
management.health.defaults.enabled=false
```

3. 自定义 HealthInfo

除了 Spring Boot 自动收集的这些 HealthInfo 之外，开发者也可以自定义 HealthInfo，只需要实现 HealthIndicator 接口即可：

```
@Component
public class SangHealth implements HealthIndicator {
    @Override
    public Health health() {
        if (checkNetwork()) {
            return Health.up().withDetail("msg", "网络连接正常...").build();
        }
        return Health.down().withDetail("msg", "网络断开...").build();
    }
}
```

代码解释：

- 开发者自定义类实现 HealthIndicator 接口，并实现该接口中的 health 方法。在 health 方法中，checkNetwork 是一个网络连接检查的方法，Health 中的 up 和 down 方法分别对应两种常见的响应状态，即 "up" 和 "down"。
- 默认的响应状态一共有 4 种，定义在 OrderedHealthAggregator 类中，分别是 DOWN、OUT_OF_SERVICE、UP、UNKNOWN，如果开发者想增加响应状态，可以自定义类继承自 HealthAggregator，或者在 application.properties 中通过 management.health.status.order 属性进行配置。

配置完成后，假设网络连接正常，访问 health 端点，结果如图 14-7 所示。

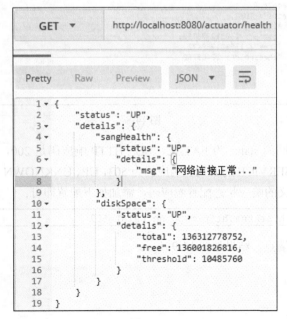

图 14-7

如果开发者想要增加响应状态 FATAL，在 application.properties 中增加如下配置：

```
1  management.health.status.order=FATAL,DOWN,OUT_OF_SERVICE,UP,UNKNOWN
```

配置完成后，就可以在 health 方法中返回自定义的响应状态了，修改 SangHealth 的 health 方法如下：

```
1  @Component
2  public class SangHealth implements HealthIndicator {
3      @Override
4      public Health health() {
5          return Health.status("FATAL").withDetail("msg", "网络断开...").build();
6      }
7  }
```

修改完成后，此时启动 Spring Boot 项目，访问 health 端点，结果如图 14-8 所示。

图 14-8

注意，此时虽然返回的 status 为 FATAL，但是 HTTP 响应码是 200，在默认的 4 种响应状态中，DOWN、OUT_OF_SERVICE 的 HTTP 响应码为 503，UP、UNKNOWN 的 HTTP 响应码为 200，如果开发者需要对自定义的响应状态配置响应码，添加如下配置即可：

```
1  management.health.status.http-mapping.FATAL=503
```

此时再访问 health 端点，结果如图 14-9 所示。

图 14-9

14.1.8 应用信息

应用信息就是通过/actuator/info 接口获取到的信息，主要包含三大类：自定义信息、Git 信息以及项目构建信息，下面分别来看。

1. 自定义信息

自定义信息可以在配置文件中添加，也可以在 Java 代码中添加。

在配置文件中添加是指在 application.properties 中手动定义以 info 开头的信息，这些信息将在 info 端点中显示出来，例如在 application.properties 文件中添加如下配置信息：

```
1  info.app.encoding=@project.build.sourceEncoding@
2  info.app.java.source=@java.version@
3  info.app.java.target=@java.version@
4  info.author.name=江南一点雨
5  info.author.email=wangsong0210@gmail.com
```

注意，@...@ 表示引用 Maven 中的定义。

添加这些配置信息后，重启 Spring Boot 项目，访问/actuator/info 端点，结果如图 14-10 所示。

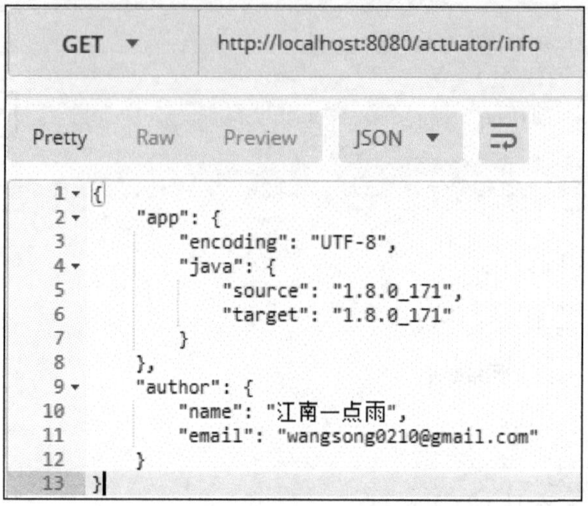

图 14-10

通过 Java 代码自定义信息只需要自定义类继承自 InfoContributor，然后实现该类中的 contribute 方法即可，代码如下：

```
@Component
public class SangInfo implements InfoContributor {
    @Override
    public void contribute(Info.Builder builder) {
        Map<String, String> info = new HashMap<>();
        info.put("name", "江南一点雨");
        info.put("email", "wangsong0210@gmail.com");
        builder.withDetail("author", info);
    }
}
```

此时访问 info 端点，结果如图 14-11 所示。

图 14-11

2．Git 信息

Git 信息是指 Git 提交信息，当 classpath 下存在一个 git.properties 文件时，Spring Boot 会自动配置一个 GitProperties Bean。开发者可通过 Git 插件自动生成 Git 提交信息，然后将这些展示在 info 端点中。具体操作步骤如下：

首先进入当前项目目录下，初始化 Git 仓库并且提交代码到本地仓库，代码如下：

```
git init
git add .
git commit -m "首次提交"
```

Git 提交完成后，接下来在 pom.xml 中添加如下 plugin：

```
<plugin>
<groupId>pl.project13.maven</groupId>
<artifactId>git-commit-id-plugin</artifactId>
</plugin>
```

使用该插件生成 Git 提交信息，插件添加成功后，接下来在 IntelliJ IDEA 中单击 Maven Project，然后找到该插件，单击 git-commit-id:revision 按钮，生成 Git 提交信息，如图 14-12 所示。

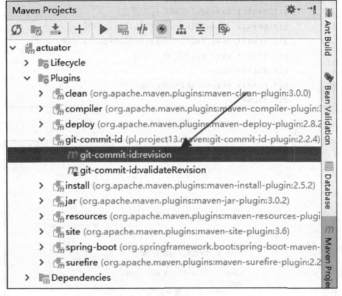

图 14-12

Git 提交信息生成成功后，在当前项目的 target/classes 目录下会看到一个 git.properties 文件，打开就是 Git 的提交信息，如图 14-13 所示。

基本上所有的 Git 提交信息都包含在这里了，如分支、提交的版本号、提交的 message、提交的用户、用户邮箱等。

最后在 application.properties 中添加如下配置，表示展示所有的 Git 提交信息：

```
management.info.git.mode=full
```

注意，management.info.git.mode 的取值还可以是 simple，表示只展示一部分核心提交信息。

配置完成后，启动 Spring Boot 项目，访问 info 端点，结果如图 14-14 所示，所有的 Git 提交信息都展示出来了。

git.properties

```
#Generated by Git-Commit-Id-Plugin
#Fri Aug 10 17:57:29 CST 2018
git.branch=master
git.build.host=DESKTOP-26A6AG4
git.build.time=2018-08-10T17\:57\:29+0800
git.build.user.email=wangsong0210@gmail.com
git.build.user.name=江南一点雨
git.build.version=0.0.1-SNAPSHOT
git.closest.tag.commit.count=
git.closest.tag.name=
git.commit.id=3a19f88f828894a095c400494761af8ca91f023d
git.commit.id.abbrev=3a19f88
git.commit.id.describe=3a19f88
git.commit.id.describe-short=3a19f88
git.commit.message.full=首次提交
git.commit.message.short=首次提交
git.commit.time=2018-08-10T17\:38\:42+0800
git.commit.user.email=wangsong0210@gmail.com
git.commit.user.name=江南一点雨
git.dirty=false
git.remote.origin.url=Unknown
git.tags=
```

图 14-13

GET http://localhost:8080/actuator/info

Pretty Raw Preview JSON


```
{
    "git": {
        "build": {
            "host": "DESKTOP-26A6AG4",
            "version": "0.0.1-SNAPSHOT",
            "time": "2018-08-10T09:57:29Z",
            "user": {
                "name": "江南一点雨",
                "email": "wangsong0210@gmail.com"
            }
        },
        "branch": "master",
        "commit": {
            "message": {
                "short": "首次提交",
                "full": "首次提交"
            },
            "id": {
                "describe": "3a19f88",
                "abbrev": "3a19f88",
                "describe-short": "3a19f88",
                "full": "3a19f88f828894a095c400494761af8ca91f023d"
            },
            "time": "2018-08-10T09:38:42Z",
            "user": {
                "email": "wangsong0210@gmail.com",
                "name": "江南一点雨"
            }
        },
        "closest": {
            "tag": {
```


图 14-14

3. 项目构建信息

如果 classpath 下存在 META-INF/build-info.properties 文件，Spring Boot 将自动构建 BuildProperties Bean，然后 info 端点会发布 build-info.properties 文件中的信息。build-info.properties 文件可以通过插件自动生成。具体操作步骤如下。

首先在 pom.xml 文件中添加插件：

```
1   <plugin>
2   <groupId>org.springframework.boot</groupId>
3   <artifactId>spring-boot-maven-plugin</artifactId>
4   <executions>
5   <execution>
6   <goals>
7   <goal>build-info</goal>
8   </goals>
9   </execution>
10  </executions>
11  </plugin>
```

然后在 IntelliJ IDEA 中单击 Maven Project，找到该插件，单击 spring-boot:build-info 按钮，生成构建信息，如图 14-15 所示。

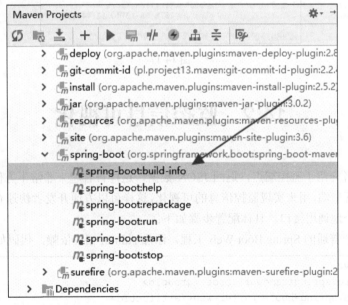

图 14-15

构建信息生成成功后，在当前项目目录下的 target/classes/META-INF 目录下生成了一个 build-info.properties 文件，内容如图 14-16 所示。

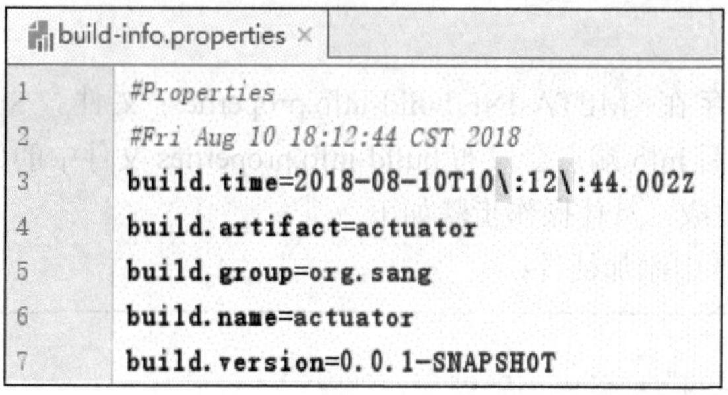

```
1  #Properties
2  #Fri Aug 10 18:12:44 CST 2018
3  build.time=2018-08-10T10\:12\:44.002Z
4  build.artifact=actuator
5  build.group=org.sang
6  build.name=actuator
7  build.version=0.0.1-SNAPSHOT
```

图 14-16

此时启动 Spring Boot 项目，访问 info 端点，构建信息将被自动发布，如图 14-17 所示。

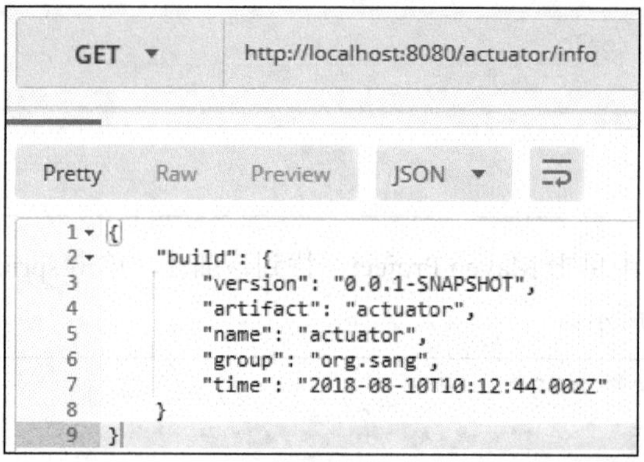

图 14-17

14.2 监控信息可视化

上一节向读者介绍了监控端点，返回 JSON 数据，这样查看起来非常不方便。Spring Boot 中提供了监控信息管理端，用来实现监控信息的可视化，这样可以方便开发者快速查看系统运行状况，而不用去一个一个地调用接口。具体配置步骤如下：

首先创建一个普通的 Spring Boot Web 工程，添加 Admin 相关依赖，代码如下：

```
1  <dependency>
2      <groupId>org.springframework.boot</groupId>
3      <artifactId>spring-boot-starter-web</artifactId>
4  </dependency>
5  <dependency>
6      <groupId>de.codecentric</groupId>
7      <artifactId>spring-boot-admin-starter-server</artifactId>
8      <version>2.0.2</version>
9  </dependency>
```

创建成功后，在项目启动类上添加@EnableAdminServer 注解，表示启动 AdminServer，代码如下：

```
1  @Spring BootApplication
2  @EnableAdminServer
3  public class AdminApplication {
4      public static void main(String[] args) {
5          SpringApplication.run(AdminApplication.class, args);
6      }
7  }
```

配置完成后，启动 Spring Boot 项目，在浏览器中输入 http://localhost:8080/index.html，结果如图 14-18 所示。

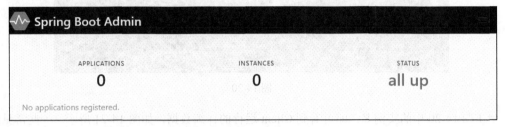

图 14-18

Admin 端将通过图表的方式展示监控信息。

接下来开发 Client。Client 实际上就是一个一个的服务，Client 将被注册到 AdminServer 上，然后 AdminServer 获取 Client 的运行数据并展示出来。因此，这里使用 14.1 节中的项目作为 Client，改造时分为以下两个步骤。

首先添加依赖：

```
1  <dependency>
2      <groupId>de.codecentric</groupId>
3      <artifactId>spring-boot-admin-starter-client</artifactId>
4      <version>2.0.2</version>
5  </dependency>
```

然后在 application.properties 中添加以下两行配置：

```
1  server.port=8081
2  spring.boot.admin.client.url=http://localhost:8080
```

spring.boot.admin.client.url 表示配置 AdminServer 地址。

配置完成后，启动 Client 项目，此时在 AdminServer 上就可以看到 Client 的运行数据。图 14-19 展示了当前注册到 AdminServer 上的 Client 列表。

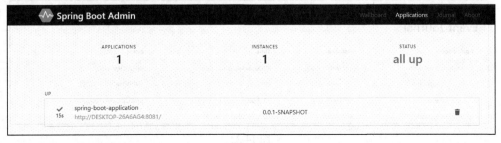

图 14-19

Wallboard 展示了 Client 的简略信息，如图 14-20 所示。

图 14-20

单击图 14-20 中的实例名，即可看到 Client 运行的详细数据，如图 14-21 所示。一些常见的信息展示在 Details 选项卡中，其他的选项卡都对应不同的端点数据。

图 14-21

Journal 中则展示了项目运行日志，如图 14-22 所示。

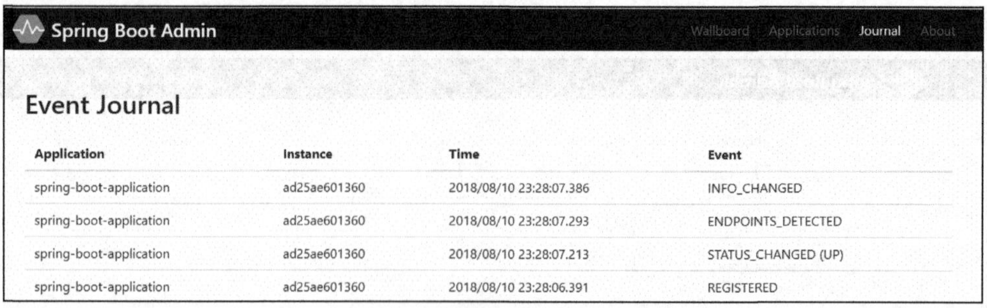

图 14-22

14.3 邮件报警

虽然使用 AdminServer 可以实现监控信息可视化，但是项目运维工程师不可能一天 24 小时盯着屏幕查看各个应用的运行状况，如果在应用运行出问题时能够自动发送邮件通知运维工程师，就会方便很多。对此，Spring Boot 提供了相应的支持。配置方式如下：

修改 14.2 节的 Admin 工程，添加邮件发送依赖：

```
1  <dependency>
2    <groupId>org.springframework.boot</groupId>
3    <artifactId>spring-boot-starter-mail</artifactId>
4  </dependency>
```

接下来在 application.properties 中配置邮件发送基本信息：

```
1   spring.mail.host=smtp.qq.com
2   spring.mail.port=465
3   spring.mail.username=xxx@qq.com
4   spring.mail.password=授权码
5   spring.mail.default-encoding=UTF-8
6   spring.mail.properties.mail.smtp.socketFactory.class=javax.net.ssl.SSLSocketFactory
7
8   spring.mail.properties.mail.debug=true
9   spring.boot.admin.notify.mail.from=xxx@qq.com
10  spring.boot.admin.notify.mail.to=xxx@qq.com
11  spring.boot.admin.notify.mail.cc=xxx@qq.com
    spring.boot.admin.notify.mail.ignore-changes=
```

关于邮件发送的配置，读者可以参考 13.1 节，第 8~11 行是新添加的配置，分别表示邮件的发送者、收件人、抄送地址以及忽略掉的事件。默认情况下，当被监控应用的状态变为 UNKNOWN 或者 UP 时不会发送报警邮件，这里的配置表示被监控应用的任何变化都会发送报警邮件。

配置完成后，重新启动 AdminServer，然后启动被监控应用，就会收到应用上线的邮件报警，如图 14-23 所示。

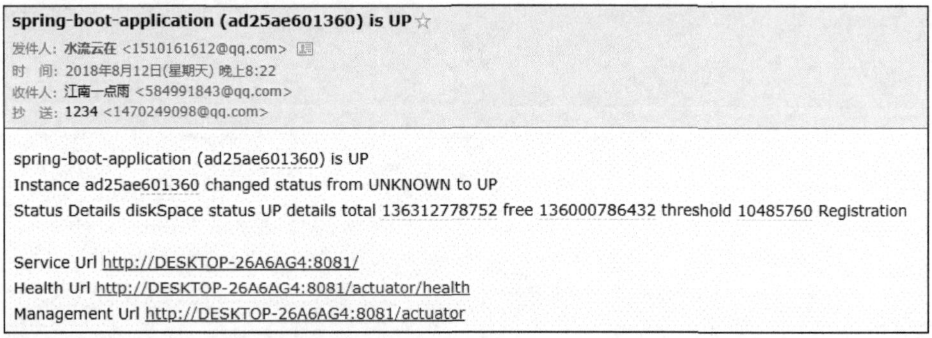

图 14-23

此时关闭被监控应用，就会收到应用下线的邮件报警，如图 14-24 所示。

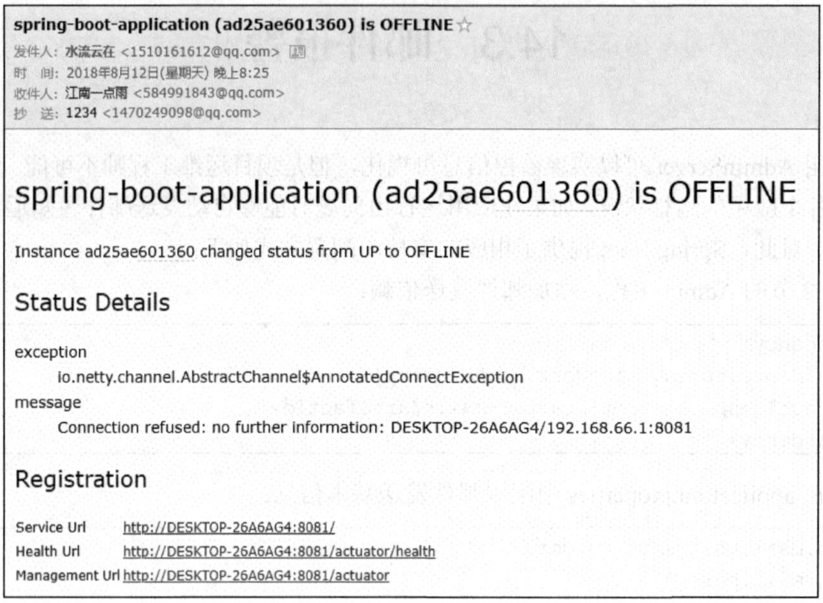

```
spring-boot-application (ad25ae601360) is OFFLINE ☆
发件人: 水流云在 <1510161612@qq.com>
时  间: 2018年8月12日(星期天) 晚上8:25
收件人: 江南一点雨 <584991843@qq.com>
抄  送: 1234 <1470249098@qq.com>
```

spring-boot-application (ad25ae601360) is OFFLINE

Instance ad25ae601360 changed status from UP to OFFLINE

Status Details

exception
　　io.netty.channel.AbstractChannel$AnnotatedConnectException
message
　　Connection refused: no further information: DESKTOP-26A6AG4/192.168.66.1:8081

Registration

Service Url　　http://DESKTOP-26A6AG4:8081/
Health Url　　http://DESKTOP-26A6AG4:8081/actuator/health
Management Url http://DESKTOP-26A6AG4:8081/actuator

图 14-24

14.4 小　　结

　　本章向读者介绍了 Spring Boot 项目中常见的应用监控，分别介绍了端点的配置以及监控数据的可视化，Spring Boot 提供的这一整套应用监控解决方案非常强大，在常规项目中稍微修改就可以直接用于生产环境了。邮件报警则可以使运维工程师及时获取应用的运行信息，特别是在应用程序下线时及时收到通知，尽早解决问题，避免造成损失。

第 15 章

项目构建与部署

本章概要

- 构建 JAR
- 构建 WAR

Spring Boot 项目可以内嵌 Servlet 容器，因此它的部署变得极为方便，可以直接打成可执行 JAR 包部署在有 Java 运行环境的服务器上，也可以像传统的 Java Web 应用程序那样打成 WAR 包运行。下面对两种构建方式分别予以介绍。

15.1 JAR

15.1.1 项目打包

使用 spring-boot-maven-plugin 插件可以创建一个可执行的 JAR 应用程序，前提是应用程序的 parent 为 spring-boot-starter-parent。配置方式如下：

```
1  <build>
2    <plugins>
3      <plugin>
4        <groupId>org.springframework.boot</groupId>
5        <artifactId>spring-boot-maven-plugin</artifactId>
6      </plugin>
7    </plugins>
```

```
8        </build>
```

配置完成后，在当前项目的根目录下执行如下 Maven 命令进行打包：

```
1        mvn package
```

或者在 IntelliJ IDEA 中单击 Maven Project，找到 Lifecycle 中的 package 双击打包，如图 15-1 所示。

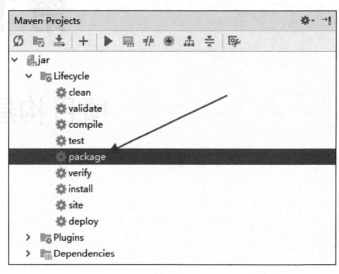

图 15-1

打包成功后，在当前项目的根目录下找到 target 目录，target 目录中就有刚刚打成的 JAR 包，如图 15-2 所示。

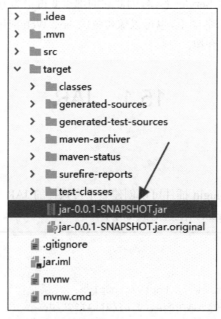

图 15-2

这种打包方式的前提是项目使用了 spring-boot-starter-parent 作为 parent，不过在大部分项目中，项目的 parent 可能并不是 spring-boot-starter-parent，而是公司内部定义好的一个配置，此时 spring-boot-maven-plugin 插件并不能直接使用，需要多做一些额外配置，代码如下：

```
1   <plugin>
2     <groupId>org.springframework.boot</groupId>
3     <artifactId>spring-boot-maven-plugin</artifactId>
4     <version>2.0.4.RELEASE</version>
5     <executions>
6       <execution>
7         <goals>
8           <goal>repackage</goal>
9         </goals>
10      </execution>
11    </executions>
12  </plugin>
```

配置完成后，就可以通过 Maven 命令或者 IntelliJ IDEA 中的 Maven 插件进行打包，打包方式和前文一致，不再赘述。

15.1.2 项目运行

Windows 中的运行比较容易，直接进入 target 目录中执行如下命令即可启动项目：

```
1   java -jar jar-0.0.1-SNAPSHOT.jar
```

在 Linux 上运行 Spring Boot 项目需要确保 Linux 上安装了 Java 运行环境。下面以 CentOS 7 为例介绍安装步骤。

首先下载 JDK，通过 xFTP 等工具上传到 Linux 上，或者直接在 Linux 上下载 JDK，下载成功后，对下载文件进行解压和重命名，命令如下：

```
1   tar -zxvf jdk-8u121-linux-x64.tar.gz
2   mv jdk1.8.0_121 java
```

然后编辑当前用户目录下的 .bash_profile 文件，配置环境变量，命令如下：

```
1   vi .bash_profile
```

在该文件中分别配置 JAVA_HOME、CLASSPATH 以及 PATH，添加完成后的文件如图 15-3 所示。

```
PATH=$PATH:$HOME/bin

export JAVA_HOME=/opt/java
export CLASSPATH=.:$JAVA_HOME/jre/lib/rt.jar:$JAVA_HOME/lib/dt.jar:$JAVA_HOME/lib/tools.jar
export PATH=$PATH:$JAVA_HOME/bin

export PATH
```

图 15-3

配置完成后，执行如下命令使配置生效：

```
1  source .bash_profile
```

接下来通过 xFTP 或者其他工具将生成的 JAR 包上传到 Linux 上，然后执行如下命令启动项目：

```
1  java -jar jar-0.0.1-SNAPSHOT.jar &
```

注意，最后面的&表示让项目在后台运行。由于在生产环境中，Linux 大多数情况下都是远程服务器，开发者通过远程工具连接 Linux，如果使用上面的命令启动 JAR，一旦窗口关闭，JAR 也就停止运行了，因此一般通过如下命令启动 JAR：

```
1  nohup java -jar jar-0.0.1-SNAPSHOT.jar &
```

这里多了 nohup，表示当窗口关闭时服务不挂起，继续在后台运行。

15.1.3 创建可依赖的 JAR

正常情况下，Spring Boot 项目是一个可以独立运行的项目，它存在的目的不是作为某一个项目的依赖，如果有一个项目需要依赖 Spring Boot 中的模块，最好的解决方案是将该模块单独拎出来，创建一个公共模块被其他项目依赖。但若由于其他原因导致该模块无法单独拎出来，此时不可以直接使用 15.1.1 小节的方法打包成 JAR 作为项目依赖，因为前面打包的 JAR 是可执行 JAR，它的类放在 BOOT-INF 目录下，如果直接作为项目的依赖，就会找不到类。可执行 JAR 的结构图如下：

```
1   example.jar
2   |
3   +-META-INF
4   |  +-MANIFEST.MF
5   +-org
6   |  +-springframework
7   |     +-boot
8   |        +-loader
9   |           +-<spring boot loader classes>
10  +-BOOT-INF
11     +-classes
12     |  +-mycompany
13     |     +-project
14     |        +-YourClasses.class
15     +-lib
16        +-dependency1.jar
17        +-dependency2.jar
```

因此，如果非要将一个 Spring Boot 工程作为一个项目的依赖，就需要配置 Maven 插件生成一个单独的 artifact，这个单独的 artifact 可以作为其他项目的依赖，配置方式如下（假设项目的 parent 不是 spring-boot-starter-parent）：

```
1  <plugin>
2  <groupId>org.springframework.boot</groupId>
3  <artifactId>spring-boot-maven-plugin</artifactId>
4  <version>2.0.4.RELEASE</version>
```

```
5  <executions>
6  <execution>
7  <goals>
8  <goal>repackage</goal>
9  </goals>
10 </execution>
11 </executions>
12 <configuration>
13 <classifier>exec</classifier>
14 </configuration>
15 </plugin>
```

classifier 指定了可执行 JAR 的名字，默认的 JAR 则作为可被其他程序依赖的 artifact。配置完成后，对项目进行打包，打包成功后，在 target 目录下生成了两个 JAR，如图 15-4 所示。

jar-0.0.1-SNAPSHOT.jar 是一个可被其他应用程序依赖的 JAR，jar-0.0.1-SNAPSHOT-exec.jar 则是一个可执行 JAR，对这两个 JAR 分别解压，可以看到 class 路径是不同的，如图 15-5 所示。

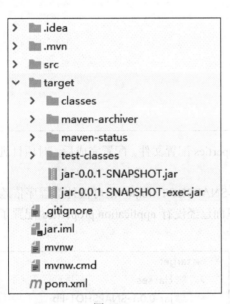

图 15-4

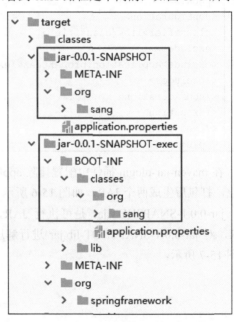

图 15-5

15.1.4 文件排除

要将 Spring Boot 项目打包成可执行 JAR，一般需要一些配置文件，例如 application.properties 或者 application.yml 等，但若将 Spring Boot 项目打包成一个可依赖 JAR，这些配置文件很多时候又不需要，此时可以在打包时排除配置文件，配置如下：

```
1  <plugin>
2  <groupId>org.springframework.boot</groupId>
3  <artifactId>spring-boot-maven-plugin</artifactId>
4  <version>2.0.4.RELEASE</version>
```

```xml
5   <executions>
6     <execution>
7       <goals>
8         <goal>repackage</goal>
9       </goals>
10    </execution>
11  </executions>
12 </plugin>
13 <plugin>
14   <artifactId>maven-jar-plugin</artifactId>
15   <executions>
16     <execution>
17       <id>lib</id>
18       <phase>package</phase>
19       <goals>
20         <goal>jar</goal>
21       </goals>
22       <configuration>
23         <classifier>lib</classifier>
24         <excludes>
25           <exclude>application.properties</exclude>
26         </excludes>
27       </configuration>
28     </execution>
29   </executions>
30 </plugin>
```

在 maven-jar-plugin 插件中配置排除 application.properties 配置文件。配置完成后，对项目进行打包，打包后生成两个 JAR，如图 15-6 所示。

jar-0.0.1-SNAPSHOT.jar 是可执行 JAR，jar-0.0.1-SNAPSHOT-lib.jar 是可被外部程序依赖的 JAR，对 jar-0.0.1-SNAPSHOT-lib.jar 进行解压，发现里面已经没有 application.properties 配置了，如图 15-7 所示。

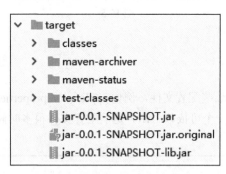

图 15-6　　　　　　　　　　　　　图 15-7

15.2 WAR

在一些特殊情况下,需要开发者将 Spring Boot 项目打成 WAR 包,然后使用传统的方式部署,打成 WAR 包的配置步骤如下:

步骤01 修改 pom.xml 文件,将项目打成 WAR 包:

```
1  <packaging>war</packaging>
```

步骤02 修改 pom.xml 文件,将内嵌容器的依赖标记为 provide,代码如下:

```
1  <dependency>
2    <groupId>org.springframework.boot</groupId>
3    <artifactId>spring-boot-starter-tomcat</artifactId>
4    <scope>provided</scope>
5  </dependency>
```

步骤03 提供一个 Spring BootServletInitializer 的子类,并覆盖其 configure 方法,完成初始化操作,代码如下:

```
1  public class ServletInitializer extends Spring BootServletInitializer {
2      @Override
3      protected SpringApplicationBuilder configure(SpringApplicationBuilder app) {
4          return app.sources(WarApplication.class);
5      }
6  }
```

经过以上三步的配置后,接下来就可以对项目进行打包了。打 WAR 包的方式和打 JAR 包的方式是一样的,打包成功后,在 target 目录下生成一个 WAR 包,将该文件复制到 Tomcat 的 webapps 目录下,启动 Tomcat 即可。

15.3 小 结

本章主要向读者介绍了 Spring Boot 项目不同的打包方式,开发者可以使用传统的 WAR 包部署,也可以使用 Spring Boot 官方推荐的 JAR 包部署,两种部署方式各有优缺点,需要开发者根据实际情况选择合适的部署方式。

第 16 章

微人事项目实战

本章概要

- 微人事项目介绍
- 项目技术架构
- 前后端分离项目构建
- 登录模块实现
- 动态加载用户菜单
- 邮件发送
- 员工资料导入导出
- 在线聊天
- 前端项目打包

本章将通过一个前后端分离项目带读者掌握目前流行的 Spring Boot+Vue 前后端分离开发环境的搭建以及项目的开发流程。本章重点向读者介绍前后端分离环境的搭建以及开发流程,也涉及少量的业务逻辑。本章项目的完整代码可以在 GitHub 上下载,下载地址为 https://github.com/lenve/vhr,本章在展示代码时仅展示项目关键步骤的核心代码。

16.1 项目简介

人事管理系统是一种常见的企业后台管理系统,它的主要目的是加强各个部门之间的协调和提高工作效率。人事管理系统提供了员工资料管理、人事管理、工资管理、统计管理以及系统管理等功能，通过人事管理系统，人事组织部门能做到以人为中心，各部门之间实现资源共享，并且实现即时通信，提高工作效率，简化烦琐的手工统计、信息汇总和工资业务等大量的人工工作，让人事组织和工资管理工作在人事组织相关的各部门之间活跃起来。

16.2 技术架构

本项目采用当下流行的前后端分离的方式开发，后端使用 Spring Boot 开发，前端使用 Vue+ElementUI 来构建 SPA。SPA 是指 Single-Page Application，即单页面应用，SPA 应用通过动态重写当前页面来与用户交互，而非传统的从服务器重新加载整个新页面。这种方法避免了页面之间切换打断用户体验，使应用程序更像一个桌面应用程序。在 SPA 中，所有的 HTML、JavaScript 和 CSS 都通过单个页面的加载来检索，或者根据用户操作动态装载适当的资源并添加到页面。在 SPA 中，前端将通过 Ajax 与后端通信。对于开发者而言，SPA 最直观的感受就是项目开发完成后，只有一个 HTML 页面，所有页面的跳转都通过路由进行导航。前后端分离的另一个好处是一个后端可以对应多个前端，由于后端只负责提供数据，前后端的交互都是通过 JSON 数据完成的，因此后端开发成功后，前端可以是 PC 端页面，也可以是 Android、iOS 以及微信小程序等。

16.2.1 Vue 简介

> Vue（读音 /vju:/，类似于 view）是一套用于构建用户界面的渐进式框架。与其他大型框架不同的是，Vue 被设计为可以自底向上逐层应用。Vue 的核心库只关注视图层，不仅易于上手，还便于与第三方库或既有项目整合。另一方面，当与现代化的工具链以及各种支持类库结合使用时，Vue 完全能够为复杂的单页应用提供驱动。
>
> ——Vue 官网

对于 Vue 的基础知识，本书不做过多介绍，由于 Vue 的文档都是中文文档，因此强烈建议初学者通读官方文档来了解 Vue 的基本使用方法（地址为 https://cn.vuejs.org/v2/guide/），本书后面将直接介绍 Vue 在项目中的使用。

16.2.2 Element 简介

Vue 桌面端组件库非常多，比较流行的有 Element、Vux、iView、mint-ui、muse-ui 等，本项目采用 Element 作为前端页面组件库。要说设计，这些 UI 库差异都不是很大，基本上都是 Material Design 风格的，本项目采用 Element 主要考虑到该库的使用人数较多（截至写作本书时，Element 在 GitHub 上的 star 数已达 29 000，接近 30 000），出了问题容易找到解决方案。关于 Element 的用法，强烈建议初学者通读官方文档学习（地址为 http://element-cn.eleme.io/#/zh-CN/component）。

16.2.3 其他

除了前端技术点外，后端用到的技术主要就是第 1~15 章提到的技术，这里就不详细展开了。

16.3 项目构建

16.3.1 前端项目构建

Vue 项目使用 webpack 来构建。首先确保本地已经安装了 NodeJS，然后在 CMD 中执行如下命令，可以创建并启动一个名为 vuehr 的前端项目：

```
1  npm install -g vue-cli
2  vue init webpack vuehr
3  cd vuehr
4  npm run dev
```

在执行"vue init webpack vuehr"命令时，会要求依次输入项目的基本信息，如图 16-1 所示。

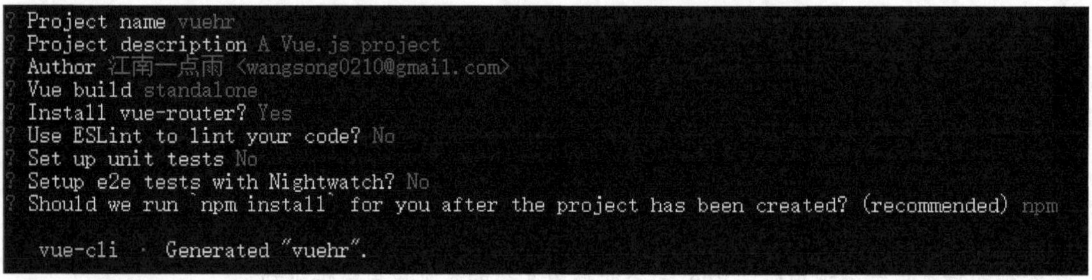

图 16-1

基本信息主要包括：

- 项目名称。
- 项目描述。
- 项目作者。
- Vue 项目构建：运行+编译还是仅运行。
- 是否安装 vue-router。
- 是否使用 ESLint。
- 是否使用单元测试。
- 是否适用 Nightwatch e2e 测试。
- 是否在项目创建成功后自动执行"npm install"安装依赖，若选择否，则在第 4 行命令执行之前执行"npm install"。

当"npm run dev"命令执行之后，在浏览器中输入 http://localhost:8080，显示页面如图 16-2 所示。

图 16-2

16.3.2 后端项目构建

后端使用 Spring Boot 创建一个 Spring Boot 工程，添加 spring-boot-starter-web 依赖即可：

```
1  <dependency>
2      <groupId>org.springframework.boot</groupId>
3      <artifactId>spring-boot-starter-web</artifactId>
4  </dependency>
```

当然，后端所需的依赖不止 spring-boot-starter-web，在后文功能不断完善的过程中，再继续添加其他依赖。另外，后端项目所需的 Redis 配置、邮件发送配置、POI 配置、WebSocket 配置等，将在涉及相关功能时向读者介绍。

16.3.3 数据模型设计

完整的数据库脚本可以在 GitHub 上下载，下载地址为 https://github.com/lenve/vhr/blob/master/hrserver/src/main/resources/vhr.sql，这里仅展示本项目的数据字典。

adjustsalary 表（员工调薪表）如表 16-1 所示。

表 16-1　adjustsalary 表

字段名	逻辑名	数据类型	约束	说明
id		Integer	主键，自增长	主键
eid		Integer	外键，普通索引	员工 id
asDate		Date		调薪日期
beforeSalary		Integer		调前薪资
afterSalary		Integer		调后薪资
reason		String(255)		调薪原因
remark		String(255)		备注

appraise 表（员工评价表）如表 16-2 所示。

表 16-2　appraise 表

字段名	逻辑名	数据类型	约束	说明
id		Integer	主键，自增长	主键
eid		Integer	外键，普通索引	员工 id
appDate		Date		考评日期
appResult		String(32)		考评结果
appContent		String(255)		考评内容
remark		String(255)		备注

department 表（部门表）如表 16-3 所示。

表 16-3　department 表

字段名	逻辑名	数据类型	约束	说明
id		Integer	主键，自增长	主键
name		String(32)		部门名称
parentId		Integer		父部门 id
depPath		String(255)		部门 path
enabled		Enum	默认值：1	是否可用
isParent		Enum	默认值：0	是否为父部门

employee 表（员工信息表）如表 16-4 所示。

表 16-4　employee 表

字段名	逻辑名	数据类型	约束	说明
id		Integer	主键，自增长	员工编号
name		String(10)		员工姓名
gender		String(4)		性别
birthday		Date		出生日期
idCard		String(18)		身份证号
wedlock		String(2)		婚姻状况
nationId		Integer(8)	外键，普通索引	民族
nativePlace		String(20)		籍贯
politicId		Integer(8)	外键，普通索引	政治面貌
email		String(20)		邮箱
phone		String(11)		电话号码
address		String(64)		联系地址
departmentId		Integer	外键，普通索引	所属部门
jobLevelId		Integer	外键，普通索引	职称 ID
posId		Integer	外键，普通索引	职位 ID
engageForm		String(8)		聘用形式
tiptopDegree		String(2)		最高学历

（续表）

字段名	逻辑名	数据类型	约束	说明
specialty		String(32)		所属专业
school		String(32)		毕业院校
beginDate		Date		入职日期
workState		String(2)	默认值：在职	在职状态
worked		String(8)	普通索引	工号
contractTerm		Float		合同期限
conversionTime		Date		转正日期
notWorkDate		Date		离职日期
beginContract		Date		合同起始日期
endContract		Date		合同终止日期
workAge		Integer		工龄

employeeec 表（员工奖惩表）如表 16-5 所示。

表 16-5 employeeec 表

字段名	逻辑名	数据类型	约束	说明
id		Integer	主键，自增长	主键
eid		Integer	外键，普通索引	员工编号
ecDate		Date		奖罚日期
ecReason		String(255)		奖罚原因
ecPoint		Integer		奖罚分
ecType		Integer		奖罚类别，0：奖，1：罚
remark		String(255)		备注

employeeremove 表（员工调岗表）如表 16-6 所示。

表 16-6 employeeremove 表

字段名	逻辑名	数据类型	约束	说明
id		Integer	主键，自增长	主键
eid		Integer	外键，普通索引	员工 id
afterDepId		Integer		调动后部门
afterJobId		Integer		调动后职位
removeDate		Date		调动日期
reason		String(255)		调动原因
remark		String(255)		备注

employeetrain 表（员工培训表）如表 16-7 所示。

表 16-7 employeetrain 表

字段名	逻辑名	数据类型	约束	说明
id		Integer	主键，自增长	主键

（续表）

字段名	逻辑名	数据类型	约束	说明
eid		Integer	外键，普通索引	员工编号
trainDate		Date		培训日期
trainContent		String(255)		培训内容
remark		String(255)		备注

empsalary 表（员工薪资关联表）如表 16-8 所示。

表 16-8　empsalary 表

字段名	逻辑名	数据类型	约束	说明
id		Integer	主键，自增长	主键
eid		Integer	外键，普通索引	员工 id
sid		Integer	外键，普通索引	薪资 id

hr 表（hr 表）如表 16-9 所示。

表 16-9　hr 表

字段名	逻辑名	数据类型	约束	说明
id		Integer	主键，自增长	hrID
name		String(32)		姓名
phone		String(11)		手机号码
telephone		String(16)		住宅电话
address		String(64)		联系地址
enabled		Enum	默认值：1	账户是否可用
username		String(255)		用户名
password		String(255)		密码
userface		String(255)		用户头像
remark		String(255)		备注

hr_role 表（hr 角色表）如表 16-10 所示。

表 16-10　hr_role 表

字段名	逻辑名	数据类型	约束	说明
id		Integer	主键，自增长	主键 id
hrid		Integer	外键，普通索引	操作员 id
rid		Integer	外键，普通索引	角色 id

joblevel 表（职称表）如表 16-11 所示。

表 16-11　joblevel 表

字段名	逻辑名	数据类型	约束	说明
id		Integer	主键，自增长	主键
name		String(32)		职称名称

（续表）

字段名	逻辑名	数据类型	约束	说明
titleLevel		String(3)		职称级别
createDate		Date	默认值：CURRENT_TIMESTAMP	创建日期
enabled		Enum	默认值：1	是否可用

menu 表（菜单表）如表 16-12 所示。

表 16-12　menu 表

字段名	逻辑名	数据类型	约束	说明
id		Integer	主键，自增长	主键
url		String(64)		请求路径规则
path		String(64)		路由 path
component		String(64)		组件名称
name		String(64)		组件名
iconCls		String(64)		菜单图标
keepAlive		Enum		菜单切换时是否保活
requireAuth		Enum		是否登录后才能访问
parentId		Integer	外键，普通索引	父菜单 id
enabled		Enum	默认值：1	是否可用

menu_role 表（菜单角色关联表）如表 16-13 所示。

表 16-13　menu_role 表

字段名	逻辑名	数据类型	约束	说明
id		Integer	主键，自增长	主键
mid		Integer	外键，普通索引	菜单 id
rid		Integer	外键，普通索引	角色 id

msgcontent 表（消息内容表）如表 16-14 所示。

表 16-14　msgcontent 表

字段名	逻辑名	数据类型	约束	说明
id		Integer	主键，自增长	主键
title		String(64)		消息标题
message		String(255)		消息内容
createDate		Date	非空，默认值：CURRENT_TIMESTAMP	创建日期

nation 表（民族表）如表 16-15 所示。

表 16-15 nation 表

字段名	逻辑名	数据类型	约束	说明
id		Integer	主键，自增长	主键
name		String(32)		名称

oplog 表（操作日志表）如表 16-16 所示。

表 16-16 oplog 表

字段名	逻辑名	数据类型	约束	说明
id		Integer	主键，自增长	主键
addDate		Date		添加日期
operate		String(255)		操作内容
hrid		Integer	外键，普通索引	操作员 ID

politicsstatus 表（政治面貌表）如表 16-17 所示。

表 16-17 politicsstatus 表

字段名	逻辑名	数据类型	约束	说明
id		Integer	主键，自增长	主键
name		String(32)		名称

position 表（职位表）如表 16-18 所示。

表 16-18 position 表

字段名	逻辑名	数据类型	约束	说明
id		Integer	主键，自增长	主键
Name		String(32)	唯一索引	职位
createDate		Date	默认值：CURRENT_TIMESTAMP	创建日期
enabled		Enum	默认值：1	是否可用

role 表（角色表）如表 16-19 所示。

表 16-19 role 表

字段名	逻辑名	数据类型	约束	说明
id		Integer	主键，自增长	主键
name		String(64)		角色名称
nameZh		String(64)		角色中文名称

salary 表（薪水表）如表 16-20 所示。

表 16-20 salary 表

字段名	逻辑名	数据类型	约束	说明
id		Integer	主键，自增长	主键

(续表)

字段名	逻辑名	数据类型	约束	说明
basicSalary		Integer		基本工资
bonus		Integer		奖金
lunchSalary		Integer		午餐补助
trafficSalary		Integer		交通补助
allSalary		Integer		应发工资
pensionBase		Integer		养老金基数
pensionPer		Float(12,31)		养老金比率
createDate		Date		启用时间
medicalBase		Integer		医疗基数
medicalPer		Float(12,31)		医疗保险比率
accumulationFundBase		Integer		公积金基数
accumulationFundPer		Float(12,31)		公积金比率
name		String(32)		账套名称

sysmsg 表（系统消息表）如表 16-21 所示。

表 16-21 sysmsg 表

字段名	逻辑名	数据类型	约束	说明
id		Integer	主键，自增长	主键
mid		Integer	外键，普通索引	消息 id
type		Integer	默认值：0	0 表示群发消息
hrid		Integer	外键，普通索引	这条消息是给谁的
state		Integer	默认值：0	0：未读，1：已读

经过以上准备工作，项目环境就已经基本搭建成功了。另外，对于 Redis 的安装、启动等，读者可以参考第 6 章，这里不再赘述。

16.4 登录模块

16.4.1 后端接口实现

后端权限认证采用 Spring Security 实现（本小节中大量知识点与第 10 章的内容相关，需要读者熟练掌握第 10 章的内容），数据库访问使用 MyBatis，同时使用 Redis 实现认证信息缓存。因此，后端首先添加如下依赖（依次是 MyBatis 依赖、Spring Security 依赖、Redis 依赖、数据库连接池依赖、数据库驱动依赖以及缓存依赖）：

```
1  <dependency>
2      <groupId>org.mybatis.spring.boot</groupId>
3      <artifactId>mybatis-spring-boot-starter</artifactId>
```

```xml
4            <version>1.3.2</version>
5        </dependency>
6        <dependency>
7            <groupId>org.springframework.boot</groupId>
8            <artifactId>spring-boot-starter-security</artifactId>
9        </dependency>
10       <dependency>
11           <groupId>org.springframework.boot</groupId>
12           <artifactId>spring-boot-starter-data-redis</artifactId>
13           <exclusions>
14               <exclusion>
15                   <groupId>io.lettuce</groupId>
16                   <artifactId>lettuce-core</artifactId>
17               </exclusion>
18           </exclusions>
19       </dependency>
20       <dependency>
21           <groupId>redis.clients</groupId>
22           <artifactId>jedis</artifactId>
23       </dependency>
24       <dependency>
25           <groupId>com.alibaba</groupId>
26           <artifactId>druid</artifactId>
27           <version>1.1.10</version>
28       </dependency>
29       <dependency>
30           <groupId>mysql</groupId>
31           <artifactId>mysql-connector-java</artifactId>
32       </dependency>
33       <dependency>
34           <groupId>org.springframework.boot</groupId>
35           <artifactId>spring-boot-starter-cache</artifactId>
36       </dependency>
```

依赖添加完成后，接下来在 application.properties 中配置数据库连接、Redis 连接以及缓存等。

```
1    #MySQL 配置
2    spring.datasource.type=com.alibaba.druid.pool.DruidDataSource
3    spring.datasource.url=jdbc:mysql://127.0.0.1:3306/vhr
4    spring.datasource.username=root
5    spring.datasource.password=root
6    #MyBatis 日志配置
7    mybatis.config-location=classpath:/mybatis-config.xml
8    #Redis 配置
9    spring.redis.database=0
10   spring.redis.host=192.168.66.130
11   spring.redis.port=6379
12   spring.redis.password=123@456
13   spring.redis.jedis.pool.max-active=8
14   spring.redis.jedis.pool.max-idle=8
15   spring.redis.jedis.pool.max-wait=-1ms
```

```
16  spring.redis.jedis.pool.min-idle=0
17  #缓存配置
18  spring.cache.cache-names=menus_cache
19  spring.cache.redis.time-to-live=1800s
20  #端口配置
21  server.port=8082
```

配置完成后，接下来实现用户认证的配置。用户认证使用 Spring Security 实现，因此需要首先提供一个 UserDetails 的实例，在人事管理系统中，登录操作是 Hr 登录，根据前面的 Hr 表创建 Hr 实体类并实现 UserDetails 接口，代码如下：

```
1   public class Hr implements UserDetails {
2       private Long id;
3       private String name;
4       private String phone;
5       private String telephone;
6       private String address;
7       private boolean enabled;
8       private String username;
9       private String password;
10      private String remark;
11      private List<Role> roles;
12      private String userface;
13      @Override
14      public boolean isEnabled() {
15          return enabled;
16      }
17      @Override
18      public String getUsername() {
19          return username;
20      }
21      @JsonIgnore
22      @Override
23      public boolean isAccountNonExpired() {
24          return true;
25      }
26      @JsonIgnore
27      @Override
28      public boolean isAccountNonLocked() {
29          return true;
30      }
31      @JsonIgnore
32      @Override
33      public boolean isCredentialsNonExpired() {
34          return true;
35      }
36      @JsonIgnore
37      @Override
38      public Collection<? extends GrantedAuthority> getAuthorities() {
39          List<GrantedAuthority> authorities = new ArrayList<>();
```

```
40              for (Role role : roles) {
41                  authorities.add(new SimpleGrantedAuthority(role.getName()));
42              }
43              return authorities;
44          }
45          @JsonIgnore
46          @Override
47          public String getPassword() {
48              return password;
49          }
50          //省略 getter/setter
51      }
```

代码解释：

- 自定义类继承自 UserDetails，并实现该接口中相关的方法。前端用户在登录成功后，需要获取当前登录用户的信息，对于一些敏感信息不必返回，使用@JsonIgnore 注解即可。
- 对于 isAccountNonExpired、isAccountNonLocked、isCredentialsNonExpired，由于 Hr 表并未设计相关字段，因此这里直接返回 true，isEnabled 方法则根据实际情况返回。
- roles 属性中存储了当前用户的所有角色信息，在 getAuthorities 方法中，将这些角色转换为 List<GrantedAuthority>的实例返回。

接下来提供一个 UserDetailsService 实例用来查询用户，代码如下：

```
1   @Service
2   public class HrService implements UserDetailsService {
3       @Autowired
4       HrMapper hrMapper;
5       @Override
6       public UserDetails loadUserByUsername(String s) throws UsernameNotFoundException{
7           Hr hr = hrMapper.loadUserByUsername(s);
8           if (hr == null) {
9               throw new UsernameNotFoundException("用户名不存在");
10          }
11          return hr;
12      }
13  }
```

自定义 HrService 实现 UserDetailsService 接口，并实现该接口中的 loadUserByUsername 方法，loadUserByUsername 方法是根据用户名查询用户的所有信息，包括用户的角色，如果没有查到相关用户，就抛出 UsernameNotFoundException 异常，表示用户不存在，如果查到了，就直接返回，由 Spring Security 框架完成密码的比对操作。

接下来需要实现动态配置权限，因此还需要提供 FilterInvocationSecurityMetadataSource 和 AccessDecisionManager 的实例。

FilterInvocationSecurityMetadataSource 代码如下：

```
1   @Component
2   public class CustomMetadataSource implements FilterInvocationSecurityMetadataSource
3   {
```

```
4       @Autowired
5       MenuService menuService;
6       AntPathMatcher antPathMatcher = new AntPathMatcher();
7       @Override
8       public Collection<ConfigAttribute> getAttributes(Object o) {
9           String requestUrl = ((FilterInvocation) o).getRequestUrl();
10          List<Menu> allMenu = menuService.getAllMenu();
11          for (Menu menu : allMenu) {
12              if (antPathMatcher.match(menu.getUrl(), requestUrl)
13  &&menu.getRoles().size()>0) {
14                  List<Role> roles = menu.getRoles();
15                  int size = roles.size();
16                  String[] values = new String[size];
17                  for (int i = 0; i < size; i++) {
18                      values[i] = roles.get(i).getName();
19                  }
20                  return SecurityConfig.createList(values);
21              }
22          }
23          return SecurityConfig.createList("ROLE_LOGIN");
24      }
25      @Override
26      public Collection<ConfigAttribute> getAllConfigAttributes() {
27          return null;
28      }
29      @Override
30      public boolean supports(Class<?> aClass) {
31          return FilterInvocation.class.isAssignableFrom(aClass);
32      }
}
```

代码解释：

- 在 getAttributes 方法中首先提取出请求 URL，根据请求 URL 判断该请求需要的角色信息。
- 通过 MenuService 中的 getAllMenu 方法获取所有的菜单资源进行比对，考虑到 getAttributes 方法在每一次请求中都会调用，因此可以将 getAllMenu 方法的返回值缓存下来，下一次请求时直接从缓存中获取。
- 对于所有未匹配成功的请求，默认都是登录后访问。

AccessDecisionManager 代码如下：

```
1   @Component
2   public class UrlAccessDecisionManager implements AccessDecisionManager {
3       @Override
4       public void decide(Authentication auth, Object o, Collection<ConfigAttribute>
5   cas) {
6           Iterator<ConfigAttribute> iterator = cas.iterator();
7           while (iterator.hasNext()) {
8               ConfigAttribute ca = iterator.next();
9               String needRole = ca.getAttribute();
```

```
10          if ("ROLE_LOGIN".equals(needRole)) {
11              if (auth instanceof AnonymousAuthenticationToken) {
12                  throw new BadCredentialsException("未登录");
13              } else
14                  return;
15          }
16          Collection<? extends GrantedAuthority> authorities =
17 auth.getAuthorities();
18          for (GrantedAuthority authority : authorities) {
19              if (authority.getAuthority().equals(needRole)) {
20                  return;
21              }
22          }
23      }
24      throw new AccessDeniedException("权限不足!");
25  }
26  @Override
27  public boolean supports(ConfigAttribute configAttribute) {
28      return true;
29  }
30  @Override
31  public boolean supports(Class<?> aClass) {
32      return true;
    }
}
```

代码解释:

- 在 decide 方法中判断当前用户是否具备请求需要的角色,若该方法在执行过程中未抛出异常,则说明请求可以通过;若抛出异常,则说明请求权限不足。
- 如果所需要的角色是 ROLE_LOGIN,那么只需要判断 auth 不是匿名用户的实例,即表示当前用户已登录。

接下来提供一个 AccessDeniedHandler 的实例来返回授权失败的信息:

```
1  @Component
2  public class AuthenticationAccessDeniedHandler implements AccessDeniedHandler {
3      @Override
4      public void handle(HttpServletRequest httpServletRequest, HttpServletResponse
5  resp,
6                         AccessDeniedException e) throws IOException {
7          resp.setStatus(HttpServletResponse.SC_FORBIDDEN);
8          resp.setContentType("application/json;charset=UTF-8");
9          PrintWriter out = resp.getWriter();
10         RespBean error = RespBean.error("权限不足,请联系管理员!");
11         out.write(new ObjectMapper().writeValueAsString(error));
12         out.flush();
13         out.close();
14     }
}
```

当授权失败时,在这里返回授权失败的信息。

当所有准备工作完成后,接下来配置 Spring Security,代码如下:

```
1   @Configuration
2   @EnableGlobalMethodSecurity(prePostEnabled = true)
3   public class WebSecurityConfig extends WebSecurityConfigurerAdapter {
4       @Autowired
5       HrService hrService;
6       @Autowired
7       CustomMetadataSource metadataSource;
8       @Autowired
9       UrlAccessDecisionManager urlAccessDecisionManager;
10      @Autowired
11      AuthenticationAccessDeniedHandler deniedHandler;
12      @Override
13      protected void configure(AuthenticationManagerBuilder auth) throws Exception {
14          auth.userDetailsService(hrService)
15                  .passwordEncoder(new BCryptPasswordEncoder());
16      }
17      @Override
18      public void configure(WebSecurity web) throws Exception {
19          web.ignoring().antMatchers("/index.html", "/static/**", "/login_p");
20      }
21      @Override
22      protected void configure(HttpSecurity http) throws Exception {
23          http.authorizeRequests()
24                  .withObjectPostProcessor(new
25  ObjectPostProcessor<FilterSecurityInterceptor>() {
26                      @Override
27                      public <O extends FilterSecurityInterceptor> O postProcess(O o) {
28                          o.setSecurityMetadataSource(metadataSource);
29                          o.setAccessDecisionManager(urlAccessDecisionManager);
30                          return o;
31                      }
32                  })
33                  .and()
34                  .formLogin().loginPage("/login_p").loginProcessingUrl("/login")
35                  .usernameParameter("username").passwordParameter("password")
36                  .failureHandler(new AuthenticationFailureHandler() {
37                      @Override
38                      public void onAuthenticationFailure(HttpServletRequest req,
39                              HttpServletResponse resp,
40                              AuthenticationException e) throws IOException {
41                          resp.setContentType("application/json;charset=utf-8");
42                          RespBean respBean = null;
43                          if (e instanceof BadCredentialsException ||
44                                  e instanceof UsernameNotFoundException) {
45                              respBean = RespBean.error("账户名或者密码输入错误!");
46                          } else if (e instanceof LockedException) {
47                              respBean = RespBean.error("账户被锁定,请联系管理员!");
```

```
48              } else if (e instanceof CredentialsExpiredException) {
49                  respBean = RespBean.error("密码过期,请联系管理员!");
50              } else if (e instanceof AccountExpiredException) {
51                  respBean = RespBean.error("账户过期,请联系管理员!");
52              } else if (e instanceof DisabledException) {
53                  respBean = RespBean.error("账户被禁用,请联系管理员!");
54              } else {
55                  respBean = RespBean.error("登录失败!");
56              }
57              resp.setStatus(401);
58              ObjectMapper om = new ObjectMapper();
59              PrintWriter out = resp.getWriter();
60              out.write(om.writeValueAsString(respBean));
61              out.flush();
62              out.close();
63          }
64      })
65      .successHandler(new AuthenticationSuccessHandler() {
66          @Override
67          public void onAuthenticationSuccess(HttpServletRequest req,
68                              HttpServletResponse resp,
69                              Authentication auth) throws IOException {
70              resp.setContentType("application/json;charset=utf-8");
71              RespBean respBean = RespBean.ok("登录成功!", HrUtils.getCurrentHr());
72              ObjectMapper om = new ObjectMapper();
73              PrintWriter out = resp.getWriter();
74              out.write(om.writeValueAsString(respBean));
75              out.flush();
76              out.close();
77          }
78      })
79      .permitAll()
80      .and()
81      .logout().permitAll()
82      .and().csrf().disable()
83      .exceptionHandling().accessDeniedHandler(deniedHandler);
84  }
}
```

代码解释:

- 首先通过@EnableGlobalMethodSecurity 注解开启基于注解的安全配置,启用@PreAuthorize 和@PostAuthorize 两个注解。
- 在配置类中注入之前创建的 4 个 Bean,在 AuthenticationManagerBuilder 中配置 userDetailsService 和 passwordEncoder。
- 在 WebSecurity 中配置需要忽略的路径。
- 在 HttpSecurity 中配置拦截规则、表单登录、登录成功或失败的响应等。
- 最后通过 accessDeniedHandler 配置异常处理。

另外,前文提到 MenuService 中的 getAllMenu 方法在每次请求时都需要查询数据库,效率极低,因此可以将该数据缓存下来,代码如下:

```
@Service
@Transactional
@CacheConfig(cacheNames = "menus_cache")
public class MenuService {
    @Autowired
    MenuMapper menuMapper;
    @Cacheable(key = "#root.methodName")
    public List<Menu> getAllMenu(){
        return menuMapper.getAllMenu();
    }
}
```

> **注 意**
>
> 这里使用方法名作为缓存的 key,另外需要在项目启动类上添加@EnableCaching 注解开启缓存。

经过前面这一整套的配置后,登录认证接口已经搭建成功了,接下来可以使用 Postman 等工具进行测试了。

登录测试如图 16-3 所示。

图 16-3

登录成功后,访问 http://localhost:8082/employee/advanced/hello 接口,由于当前用户不具备相应的角色,访问结果如图 16-4 所示。

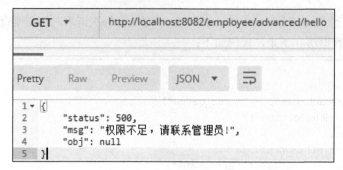

图 16-4

若访问 http://localhost:8082/employee/basic/hello 地址,则可以看到正常的结果,如图 16-5 所示。

图 16-5

确认后端接口均可以正常运行后,接下来开发前端。

16.4.2 前端实现

1. 引入 Element 和 Axios

前端 UI 使用 Element,网络请求则使用 Axios,因此首先安装 Element 和 Axios 依赖,代码如下:

```
1  npm i element-ui -S
2  npm i axios -S
```

依赖添加成功后,接下来在 main.js 中引入 Element,代码如下:

```
1  import ElementUI from 'element-ui'
2  import 'element-ui/lib/theme-chalk/index.css'
3  Vue.use(ElementUI)
```

引入 Element 之后,接下来就可以在项目中直接使用相关组件了。

对于网络请求,由于在每一次请求时都需要判断各种异常情况,然后提示用户,例如请求是否成功、失败的原因等,考虑到这些判断基本上都使用重复的代码,因此可以将网络请求封装,做成 Vue 的插件方便使用。由于封装的代码比较长,这里就不贴出来了,读者可以在 GitHub 上查看,地址为 https://github.com/lenve/vhr/blob/master/vuehr/src/utils/api.js。配置完成后,在 main.js 中导入封装的方法,然后配置为 Vue 的 prototype,代码如下:

```
1  import {getRequest} from './utils/api'
2  import {postRequest} from './utils/api'
```

```
3   import {deleteRequest} from './utils/api'
4   import {putRequest} from './utils/api'
5   Vue.prototype.getRequest = getRequest;
6   Vue.prototype.postRequest = postRequest;
7   Vue.prototype.deleteRequest = deleteRequest;
8   Vue.prototype.putRequest = putRequest;
```

配置完成后，接下来对于任何需要使用网络请求的地址，都可以使用 this.XXX 执行一个网络请求，例如要执行登录请求，就可以通过 this.postRequest(url,param) 执行。

2. 开发 Login 页面

接下来在 components 目录下创建 Login.vue 页面进行登录页面开发，代码如下：

```
1    <template>
2    <el-form :rules="rules" class="login-container" label-position="left"
3            label-width="0px" v-loading="loading">
4    <h3 class="login_title">系统登录</h3>
5    <el-form-item prop="account">
6    <el-input type="text" v-model="loginForm.username"
7            auto-complete="off" placeholder="账号"></el-input>
8    </el-form-item>
9    <el-form-item prop="checkPass">
10   <el-input type="password" v-model="loginForm.password"
11           auto-complete="off" placeholder="密码"></el-input>
12   </el-form-item>
13   <el-checkbox class="login_remember" v-model="checked"
14           label-position="left">记住密码</el-checkbox>
15   <el-form-item style="width: 100%">
16   <el-button type="primary" style="width: 100%"
17   @click="submitClick">登录</el-button>
18   </el-form-item>
19   </el-form>
20   </template>
21   <script>
22     export default{
23       data(){
24         return {
25           rules: {
26             account: [{required: true, message: '请输入用户名', trigger: 'blur'}],
27             checkPass: [{required: true, message: '请输入密码', trigger: 'blur'}]
28           },
29           checked: true,
30           loginForm: {
31             username: 'admin',
32             password: '123'
33           },
34           loading: false
35         }
36       },
37       methods: {
```

```
        submitClick: function () {
          var _this = this;
          this.loading = true;
          this.postRequest('/login', {
            username: this.loginForm.username,
            password: this.loginForm.password
          }).then(resp=> {
            _this.loading = false;
            if (resp && resp.status == 200) {
              var data = resp.data;
              _this.$store.commit('login', data.obj);
              var path = _this.$route.query.redirect;
              _this.$router
.replace({path: path == '/' || path == undefined ? '/home' : path});
            }
          });
        }
      }
    }
</script>
<style>
  .login-container {
    border-radius: 15px;
    background-clip: padding-box;
    margin: 180px auto;
    width: 350px;
    padding: 35px 35px 15px 35px;
    background: #fff;
    border: 1px solid #eaeaea;
    box-shadow: 0 0 25px #cac6c6;
  }
  .login_title {
    margin: 0px auto 40px auto;
    text-align: center;
    color: #505458;
  }
  .login_remember {
    margin: 0px 0px 35px 0px;
    text-align: left;
  }
</style>
```

代码解释：

- 系统登录使用 Element 中的 el-form 来实现。同时使用了 Element 标签提供的校验规则。
- 当用户单击"登录"按钮时，通过 this.postRequest 方法发起一个登录请求，登录成功后，将登录的用户信息保存到 store 中，同时跳转到 Home 页，或者某个重定向页面。

3. 配置路由

登录页面开发完成后，接下来在路由中配置登录页面，代码如下：

```
1  import Vue from 'vue'
2  import Router from 'vue-router'
3  import Login from '@/components/Login'
4  import Home from '@/components/Home'
5  Vue.use(Router)
6
7  export default new Router({
8    routes: [
9      {
10       path: '/',
11       name: 'Login',
12       component: Login,
13       hidden: true
14     }, {
15       path: '/home',
16       name: '主页',
17       component: Home,
18       hidden: true,
19       meta: {
20         requireAuth: true
21       }
22     }
23   ]
24 })
```

另外,由于 main.js 是入口 JS,在 main.js 中导入了 App 组件,App 组件默认有 Vue 的 Logo,将 Logo 图片删除,只保留一个<router-view/>即可,修改后的 App.vue 如下:

```
1  <template>
2    <div id="app">
3      <router-view/>
4    </div>
5  </template>
```

4. 配置请求转发

最后,由于前端项目和后端项目在不同的端口下启动,前端的网络请求无法直接发送到后端,因此需要配置请求转发。下面介绍配置方式。

修改 config 目录下的 index.js 文件,修改 proxyTable,代码如下:

```
1  proxyTable: {
2    '/': {
3      target: 'http://localhost:8082',
4      changeOrigin: true,
5      pathRewrite: {
6        '^/': ''
7      }
8    },
9    '/ws/*': {
10     target: 'ws://127.0.0.1:8082',
11     ws: true
```

```
12      }
13    },
```

这里配置了两条规则，第一条是配置 HTTP 请求转发，第二条是配置 WebSocket 请求转发，WebSocket 请求在本项目的即时通信模块中会用到。

5. 启动前端项目

做完这些操作后，接下来打开 CMD 命令窗口，进入当前项目目录下，执行如下命令启动项目：

```
1   npm run dev
```

如果开发者使用 WebStorm 开发前端项目，也可以单击 WebStorm 右上角的下拉按钮（见图 16-6），然后单击+，选择 npm（见图 16-7），配置 Name 和启动脚本（见图 16-8）。

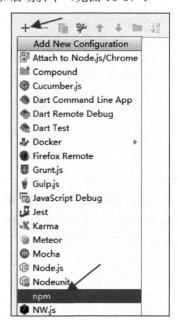

图 16-6　　　　　　　　　　　　　　　　图 16-7

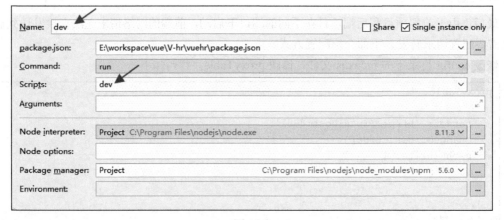

图 16-8

配置完成后，就可以直接通过单击 WebStorm 右上角的"启动"按钮启动项目了，如图 16-9 所示。

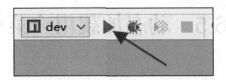

图 16-9

6. 测试

当前端项目启动成功后,接下来在浏览器中输入 http://localhost:8080,即可看到登录页面,如图 16-10 所示。

图 16-10

输入用户名和密码,单击"登录"按钮,即可登录成功,通过 Chrome 调试工具可以看到登录请求,如图 16-11 所示。

图 16-11

至此,登录功能就实现了。这里展示的只是部分核心代码,完整代码可以在 GitHub 上下载,下载地址为 https://github.com/lenve/vhr。

16.5　动态加载用户菜单

用户菜单就是用户登录成功后首页左侧显示的菜单，如图 16-12 所示。这个菜单数据是根据用户的角色动态加载的，即不同身份的用户登录成功后看到的菜单是不一样的。接下来看这个功能如何实现。

图 16-12

16.5.1　后端接口实现

后端接口的实现比较容易，根据登录用户的 id 查询该用户具有的角色，再根据角色信息查看对应的 Menu，数据模型如图 16-13 所示。

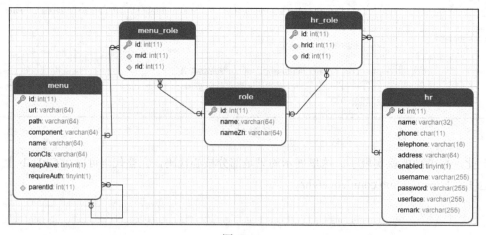

图 16-13

首先创建 MenuMapper，根据用户 id 查询 Menu，代码如下：

```
1  public interface MenuMapper {
2      List<Menu> getMenusByHrId(Long hrId);
3      //省略其他方法
4  }
```

对应的 MenuMapper.xml 文件中则根据当前用户 id 查询用户可以查看的角色，查询 SQL 如下（源文件过大，这里就不展示了，完整文件可以在 GitHub 上下载，下载地址为 https://github.com/lenve/vhr/blob/master/hrserver/src/main/java/org/sang/mapper/MenuMapper.xml）：

```
1  SELECT DISTINCT m1.*,m2.`id` AS id2,m2.`component` AS commponent2,m2.`enabled` AS
2  enabled2,m2.`keepAlive` AS keepAlive2,m2.`name` AS name2,m2.`path` AS path2,m2.`url`
3  AS url2,m2.`requireAuth` AS requireAuth2,m2.`parentId` AS parentId2 FROM menu m1,menu
4  m2,menu_role mr,role r,hr_role hrr  WHERE m1.`id`=m2.`parentId` AND mr.`rid`=r.`id`
5  AND mr.`mid`=m2.`id` AND hrr.`rid`=r.`id` AND hrr.`hrid`=#{id}
```

然后分别创建 MenuService 和 ConfigController，ConfigController 用来返回基本的系统配置信息。

MenuService 代码如下：

```
1  @Service
2  public class MenuService {
3      @Autowired
4      MenuMapper menuMapper;
5      public List<Menu> getMenusByHrId() {
6          return menuMapper.getMenusByHrId(HrUtils.getCurrentHr().getId());
7      }
8      //省略其他方法
9  }
10 public class HrUtils {
11     public static Hr getCurrentHr() {
12         return (Hr)
13 SecurityContextHolder.getContext().getAuthentication().getPrincipal();
14     }
}
```

其中，HrUtils 是一个工具方法，用来返回当前登录用户的信息。

ConfigController 代码如下：

```
1  @RestController
2  @RequestMapping("/config")
3  public class ConfigController {
4      @Autowired
5      MenuService menuService;
6      @RequestMapping("/sysmenu")
7      public List<Menu> sysmenu() {
8          return menuService.getMenusByHrId();
9      }
10 }
```

配置完成后，启动 Spring Boot 项目，访问 http://localhost:8082/config/sysmenu 接口，即可看到当前登录用户所能查看的菜单数据，如图 16-14 所示。

```
GET ▼   http://localhost:8082/config/sysmenu

Pretty   Raw   Preview   JSON ▼

 1 ▼ [
 2 ▼     {
 3           "id": 2,
 4           "path": "/home",
 5           "component": "Home",
 6           "name": "员工资料",
 7           "iconCls": "fa fa-user-circle-o",
 8           "children": [
 9 ▼             {
10                 "id": null,
11                 "path": "/emp/basic",
12                 "component": "EmpBasic",
13                 "name": "基本资料",
14                 "iconCls": null,
15                 "children": [],
16 ▼               "meta": {
17                     "keepAlive": false,
18                     "requireAuth": true
19                 }
20             }
21         ],
22 ▼       "meta": {
23             "keepAlive": false,
24             "requireAuth": true
25         }
26     },
27 ▼   {
28         "id": 3,
29         "path": "/home",
30         "component": "Home",
```

图 16-14

16.5.2　前端实现

后端返回了菜单数据，前端请求该接口获取菜单数据，这里的步骤很简单，主要分两步：

- 将服务端返回的 JSON 动态添加到当前路由中。
- 将服务端返回的 JSON 数据保存到 store 中，然后各个 Vue 页面根据 store 中的数据来渲染菜单。

这里涉及的第一个问题是请求时机，即何时去请求菜单数据。如果直接在登录成功之后请求菜单资源，那么在请求到 JSON 数据之后，将其保存在 store 中，以便下一次使用。但是这样会有一个问题：假如用户登录成功之后，单击 Home 页的某一个按钮，进入某一个子页面中，然后按一下 F5 键进行刷新，这个时候就会出现空白页面，因为按 F5 键刷新之后 store 中的数据就没了，而我们又只在登录成功的时候请求了一次菜单资源。要解决这个问题，有两种方案：方案一，不要将菜单资源保存到 store 中，而是保存到 localStorage 中，这样即使按 F5 键刷新之后数据还在；方案二，直接在每一个页面的 mounted 方法中都加载一次菜单资源。由于菜单资源是非常敏感的，因此不建议将其保存到本地，故舍弃方案一，但是方案二的工作量有点大，而且也不易维护，这里可以使用路由中的导航守卫来简化方案二的工作量。下面介绍具体实现步骤。

首先在 store 中创建一个 routes 数组，这是一个空数组，代码如下：

```
1   import Vue from 'vue'
2   import Vuex from 'vuex'
3
4   Vue.use(Vuex)
5
6   export default new Vuex.Store({
7     state: {
8       routes: []
9     },
10    mutations: {
11      initMenu(state, menus){
12        state.routes = menus;
13      }
14  });
```

然后开启路由全局守卫，代码如下：

```
1   router.beforeEach((to, from, next)=> {
2     if (to.name == 'Login') {
3       next();
4       return;
5     }
6     var name = store.state.user.name;
7     if (name == '未登录') {
8       if (to.meta.requireAuth || to.name == null) {
9         next({path: '/', query: {redirect: to.path}})
10      } else {
11        next();
12      }
13    } else {
14      initMenu(router, store);
15      next();
16    }
17  }
18  )
```

代码解释：

- 这里使用 router.beforeEach 配置了一个全局前置守卫。
- 首先判断目标页面是不是 Login，若是 Login 页面，则直接通过，因为 Login 页面不需要菜单数据。
- 接下来获取 store 中保存的当前登录的用户数据，若获取到的用户名为"未登录"，则表示用户尚未登录，在用户尚未登录的情况下，如果要跳转到某一个页面，就需要判断该页面是否要求登录后才能访问，若要求了，则直接跳转到登录页面，并配置 redirect 参数。若用户已经登录，则先执行 initMenu 方法初始化菜单数据，然后通过 next();进入下一个页面。

初始化菜单的操作如下：

```js
export const initMenu = (router, store)=> {
  if (store.state.routes.length > 0) {
    return;
  }
  getRequest("/config/sysmenu").then(resp=> {
    if (resp && resp.status == 200) {
      var fmtRoutes = formatRoutes(resp.data);
      router.addRoutes(fmtRoutes);
      store.commit('initMenu', fmtRoutes);
    }
  })
}
export const formatRoutes = (routes)=> {
  let fmRoutes = [];
  routes.forEach(router=> {
    let {
      path,
      component,
      name,
      meta,
      iconCls,
      children
    } = router;
    if (children && children instanceof Array) {
      children = formatRoutes(children);
    }
    let fmRouter = {
      path: path,
      component(resolve){
        if (component.startsWith("Home")) {
          require(['../components/' + component + '.vue'], resolve)
        } else if (component.startsWith("Emp")) {
          require(['../components/emp/' + component + '.vue'], resolve)
        } else if (component.startsWith("Per")) {
          require(['../components/personnel/' + component + '.vue'], resolve)
        } else if (component.startsWith("Sal")) {
          require(['../components/salary/' + component + '.vue'], resolve)
        } else if (component.startsWith("Sta")) {
          require(['../components/statistics/' + component + '.vue'], resolve)
        } else if (component.startsWith("Sys")) {
          require(['../components/system/' + component + '.vue'], resolve)
        }
      },
      name: name,
      iconCls: iconCls,
      meta: meta,
      children: children
    };
    fmRoutes.push(fmRouter);
  })
```

```
51      return fmRoutes;
52  }
```

代码解释：

- 在初始化菜单中，首先判断 store 中的数据是否存在，如果存在，则说明这次跳转是正常的跳转，而不是用户按 F5 键或者直接在地址栏输入某个地址进入的，这时直接返回，不必执行菜单初始化。
- 若 store 中不存在菜单数据，则需要初始化菜单数据，通过 getRequest("/config/sysmenu")方法获得菜单 JSON 数据之后，首先通过 formatRoutes 方法将服务器返回的 JSON 转为 router 需要的格式，这里主要是转 component，因为服务端返回的 component 是一个字符串，而 router 中需要的却是一个组件，因此我们在 formatRoutes 方法中根据服务端返回的 component 动态加载需要的组件即可。
- 数据格式准备成功之后，一方面将数据存到 store 中，另一方面利用路由中的 addRoutes 方法将之动态添加到路由中。

加载到路由数据之后，接下来就是菜单渲染了。菜单渲染操作在 Home.vue 组件中完成，部分核心代码如下：

```
1   <template>
2   <div>
3   <el-container class="home-container">
4   <el-header class="home-header">
5   </el-header>
6   <el-container>
7   <el-aside width="180px" class="home-aside">
8   <div>
9   <el-menu unique-opened router>
10  <template v-for="(item,index) in this.routes" v-if="!item.hidden">
11  <el-submenu :key="index" :index="index+''">
12  <template slot="title">
13  <i :class="item.iconCls"></i>
14  <span slot="title">{{item.name}}</span>
15  </template>
16  <el-menu-item width="180px"
17                  v-for="child in item.children"
18                  :index="child.path"
19                  :key="child.path">{{child.name}}
20  </el-menu-item>
21  </el-submenu>
22  </template>
23  </el-menu>
24  </div>
25  </el-aside>
26  <el-main>
27  </el-main>
28  </el-container>
29  </el-container>
```

```
30    </div>
31   </template>
32   <script>
33    export default{
34      computed: {
35        user(){
36          return this.$store.state.user;
37        },
38        routes(){
39          return this.$store.state.routes
40        }
41      }
42    }
43   </script>
```

代码解释：

- 在计算属性中返回 routes 数据。
- 遍历 routes 中的数据，根据 routes 中的数据渲染出 el-submenu 和 el-menu-item。

配置完成后，启动前端项目，使用不同的身份登录，登录成功后，就可以看到不同用户对应不同的操作菜单了。图 16-15 所示是系统管理员看到的菜单数据。图 16-16 所示是用户曾巩看到的菜单数据。

图 16-15

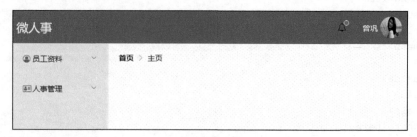

图 16-16

动态加载用户菜单就完全实现了，完整代码可以在 https://github.com/lenve/vhr 下载。

16.6 员工资料模块

完成登录模块和菜单加载模块之后，一个前后端分离的项目框架基本上就搭建成功了。接下来是业务的开发，主要是后端提供接口，前端提供页面并请求数据。下面向读者介绍员工资料模块的开发。员工资料模块页面如图 16-17 所示。

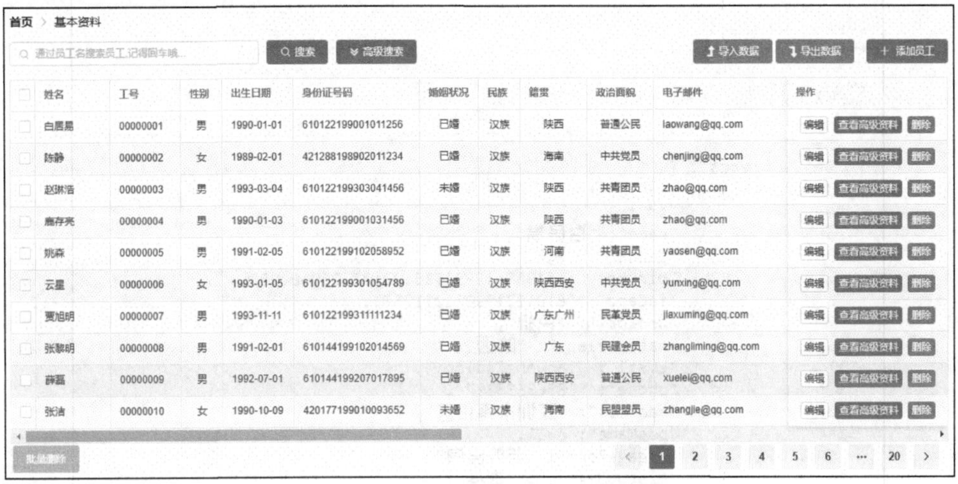

图 16-17

16.6.1 后端接口实现

员工基本资料数据的展示，后端只需要提供一个分页查询+条件查询的接口即可，代码如下：

```java
@RequestMapping(value = "/emp", method = RequestMethod.GET)
public Map<String, Object> getEmployeeByPage(
        @RequestParam(defaultValue = "1") Integer page,
        @RequestParam(defaultValue = "10") Integer size,
        @RequestParam(defaultValue = "") String keywords,
        Long politicId, Long nationId, Long posId,
        Long jobLevelId, String engageForm,
        Long departmentId, String beginDateScope) {
    Map<String, Object> map = new HashMap<>();
    List<Employee> employeeByPage = empService.getEmployeeByPage(page, size,
            keywords,politicId, nationId, posId, jobLevelId, engageForm,
            departmentId, beginDateScope);
    Long count = empService.getCountByKeywords(keywords, politicId, nationId,
            posId,jobLevelId, engageForm, departmentId, beginDateScope);
    map.put("emps", employeeByPage);
    map.put("count", count);
```

```
17        return map;
18    }
```

分页查询中，page 默认为 1，size 默认为 10，查询关键字 keywords 默认为空字符串，后面几个参数则根据政治面貌、民族、职位、职称、聘用形式、部门以及入职日期查询。具体的查询代码比较简单，这里就不贴出来了。

后端接口开发成功后，先用 Postman 进行测试，结果如图 16-18 所示。

图 16-18

> **注　意**
>
> 员工资料中的基本资料一项，接口设计规则为 "/employee/basic/**"，这是为了和数据库保持一致，防止没有权限的用户拿到请求接口后直接去请求数据。

确认后端接口可以调通后，接下来就可以开发前端页面了。

16.6.2 前端实现

前端数据的展示使用 Element 中的表格来实现。在 components 目录下创建 emp 文件夹，然后在 emp 文件夹下创建 EmpBasic.vue 用来展示前端数据。注意，这里之所以将页面放在 emp 目录下，是为了配合 16.5 节提到的菜单数据格式化（在数据格式化时，设置员工资料相关的页面都放在 emp 目录下）。

EmpBasic.vue 核心代码如下：

```
1   <template>
2   <div>
3   <el-container>
4   <el-header>
5   </el-header>
6   <el-main>
7   <el-table
8           :data="emps"
9           v-loading="tableLoading"
10          border
11          stripe
12          @selection-change="handleSelectionChange"
13          size="mini"
14          style="width: 100%">
15  <el-table-column
16          prop="gender"
17          label="性别"
18          width="50">
19  </el-table-column>
20  <el-table-column
21          width="85"
22          align="left"
23          label="出生日期">
24  <template slot-scope="scope">
25          {{ scope.row.birthday | formatDate}}
26          </template>
27  </el-table-column>
28  <el-table-column
29          prop="idCard"
30          width="150"
31          align="left"
32          label="身份证号码">
33  </el-table-column>
34
35  <!--省略部分代码-→
36
37
38  </el-table>
39  </div>
40  </el-main>
41  </el-container>
42  </div>
43  </template>
44  <script>
45    export default {
46      data() {
```

```
47        return {
48          emps: [],
49          keywords: ''
50        }
51      };
52    },
53    mounted: function () {
54      this.loadEmps();
55    },
56    methods: {
57      loadEmps(){
58        var _this = this;
59        this.tableLoading = true;
60        this.getRequest("/employee/basic/emp?page=" + this.currentPage
61          + "&size=10&keywords=" + this.keywords).then(resp=> {
62          this.tableLoading = false;
63          if (resp && resp.status == 200) {
64            var data = resp.data;
65            _this.emps = data.emps;
66            _this.totalCount = data.count;
67          }
68        })
69      }
70    }
71  };
72  </script>
```

代码解释：

- 首先在 data 中定义 emps 用来存放所有员工的 JSON 数据，然后定义 keywords 用来存放查询的关键字。
- 在加载该页面时，在 mounted 中调用 loadEmps 初始化员工数据。

另外注意，表格中的日期使用 formatDate 过滤器实现日期的格式化，formatDate 是一个全局过滤器。

formatDate 过滤器定义如下：

```
1   import Vue from 'vue'
2   Vue.filter("formatDate", formatDate);
3   function formatDate(value) {
4     var date = new Date(value);
5     var year = date.getFullYear();
6     var month = date.getMonth() + 1;
7     var day = date.getDate();
8     if (month < 10) {
9       month = "0" + month;
10    }
11    if (day < 10) {
12      day = "0" + day;
13    }
```

```
14      return year + "-" + month + "-" + day;
15  }
```

经过以上几步配置后,读者就可以看到如图 16-17 所示的效果图了。

16.7 配置邮件发送

在员工资料模块,hr 表可以手动录入员工数据,当一个员工数据被成功录入系统后,系统会自动向该员工发送一封欢迎入职邮件,要实现这个功能非常容易(关于详细的邮件发送配置,读者可以参考 13.1 节),下面介绍其具体配置。

首先在后端项目中引入邮件发送相关依赖:

```
1  <dependency>
2      <groupId>org.springframework.boot</groupId>
3      <artifactId>spring-boot-starter-mail</artifactId>
4  </dependency>
5  <dependency>
6      <groupId>org.springframework.boot</groupId>
7      <artifactId>spring-boot-starter-thymeleaf</artifactId>
8  </dependency>
```

这里除了邮件发送依赖外,还引入了 Thymeleaf 依赖,Thymeleaf 的作用是构建邮件模板。

依赖添加成功后,接下来在 application.properties 中配置邮件:

```
1  spring.mail.host=smtp.qq.com
2  spring.mail.port=465
3  spring.mail.username=xxx@qq.com
4  spring.mail.password=13.1.1 小节申请到的授权码
5  spring.mail.default-encoding=UTF-8
6  spring.mail.properties.mail.smtp.socketFactory.class=javax.net.ssl.SSLSocketFactory
7  spring.mail.properties.mail.debug=true
```

配置完成后,接下来在 resources/templates 目录下创建邮件模板 email.html,代码如下:

```
1   <!DOCTYPE html>
2   <html lang="en" xmlns:th="http://www.thymeleaf.org">
3   <head>
4   <meta charset="UTF-8">
5   <title>Title</title>
6   </head>
7   <body>
8   <p>你好,
9   <span th:text="${name}"></span>同学,欢迎加入 XXX 大家庭!您的入职信息如下:</p>
10  <table border="1" cellspacing="0">
11  <tr>
12  <td><strong style="color: #F00">工号</strong></td>
13  <td th:text="${workID}"></td>
```

```html
14  </tr>
15  <tr>
16  <td><strong style="color: #F00">合同期限</strong></td>
17  <td th:text="${contractTerm}+'年'"></td>
18  </tr>
19  <tr>
20  <td><strong style="color: #F00">合同起始日期</strong></td>
21  <td th:text="${#dates.format(beginContract, 'yyyy-MM-dd')}"></td>
22  </tr>
23  <tr>
24  <td><strong style="color: #F00">合同截至日期</strong></td>
25  <td th:text="${#dates.format(endContract, 'yyyy-MM-dd')}"></td>
26  </tr>
27  <tr>
28  <td><strong style="color: #F00">所属部门</strong></td>
29  <td th:text="${departmentName}"></td>
30  </tr>
31  <tr>
32  <td><strong style="color: #F00">职位</strong></td>
33  <td th:text="${posName}"></td>
34  </tr>
35  </table>
36  <p>
37  <strong style="color: #F00; font-size: 24px;">
38      希望在未来的日子里，携手共进！
39  </strong>
40  </p>
41  </body>
42  </html>
43
```

考虑到邮件发送是一个耗时操作，因此在子线程中完成邮件发送操作，代码如下：

```java
1   public class EmailRunnable implements Runnable {
2       private Employee employee;
3       private JavaMailSender javaMailSender;
4       private TemplateEngine templateEngine;
5
6       public EmailRunnable(Employee employee,
7                            JavaMailSender javaMailSender,
8                            TemplateEngine templateEngine) {
9           this.employee = employee;
10          this.javaMailSender = javaMailSender;
11          this.templateEngine = templateEngine;
12      }
13      @Override
14      public void run() {
15          try {
16              MimeMessage message = javaMailSender.createMimeMessage();
17              MimeMessageHelper helper = new MimeMessageHelper(message, true);
18              helper.setTo(employee.getEmail());
```

```
19              helper.setFrom("1510161612@qq.com");
20              helper.setSubject("XXX集团：通知");
21              Context ctx = new Context();
22              ctx.setVariable("name", employee.getName());
23              ctx.setVariable("workID", employee.getWorkID());
24              ctx.setVariable("contractTerm", employee.getContractTerm());
25              ctx.setVariable("beginContract", employee.getBeginContract());
26              ctx.setVariable("endContract", employee.getEndContract());
27              ctx.setVariable("departmentName", employee.getDepartmentName());
28              ctx.setVariable("posName", employee.getPosName());
29              String mail = templateEngine.process("email.html", ctx);
30              helper.setText(mail, true);
31              javaMailSender.send(message);
32          } catch (MessagingException e) {
33              System.out.println("发送失败");
34          } catch (javax.mail.MessagingException e) {
35              e.printStackTrace();
36          }
37      }
38  }
```

最后，在用户添加成功后，启动该线程发送邮件，代码如下：

```
1   @RequestMapping(value = "/emp", method = RequestMethod.POST)
2   public RespBean addEmp(Employee employee) {
3       if (empService.addEmp(employee) == 1) {
4           List<Position> allPos = positionService.getAllPos();
5           for (Position allPo : allPos) {
6               if (allPo.getId() == employee.getPosId()) {
7                   employee.setPosName(allPo.getName());
8               }
9           }
10          executorService.execute(new EmailRunnable(employee,
11                  javaMailSender, templateEngine));
12          return RespBean.ok("添加成功!");
13      }
14      return RespBean.error("添加失败!");
15  }
```

配置完成后，重启后端项目，此时在前端添加用户，添加完成后，系统会根据所添加用户的邮箱自动发送一封欢迎入职邮件，如图16-19所示。

图 16-19

完整的邮件发送代码可以在 GitHub 上查看，地址为 https://github.com/lenve/vhr。

16.8 员工资料导出

将员工资料导出为 Excel 是一个非常常见的需求，后端提供导出接口，前端下载导出数据即可。

16.8.1 后端接口实现

后端实现主要是将查询到的员工数据集合转为可以下载的 ResponseEntity<byte[]>，代码如下：

```
1   public static ResponseEntity<byte[]> exportEmp2Excel(List<Employee> emps) {
2       HttpHeaders headers = null;
3       ByteArrayOutputStream baos = null;
4       try {
5           //1.创建 Excel 文档
6           HSSFWorkbook workbook = new HSSFWorkbook();
7           //2.创建文档摘要
8           workbook.createInformationProperties();
9           //3.获取文档信息，并配置
10          DocumentSummaryInformation dsi =
11  workbook.getDocumentSummaryInformation();
12          //3.1 文档类别
13          dsi.setCategory("员工信息");
14          //3.2 设置文档管理员
15          dsi.setManager("江南一点雨");
16          //3.3 设置组织机构
17          dsi.setCompany("XXX 集团");
18          //4.获取摘要信息并配置
19          SummaryInformation si = workbook.getSummaryInformation();
20          //4.1 设置文档主题
```

```java
21    si.setSubject("员工信息表");
22    //4.2.设置文档标题
23    si.setTitle("员工信息");
24    //4.3 设置文档作者
25    si.setAuthor("XXX 集团");
26    //4.4 设置文档备注
27    si.setComments("备注信息暂无");
28    //创建 Excel 表单
29    HSSFSheet sheet = workbook.createSheet("XXX 集团员工信息表");
30    //创建日期显示格式
31    HSSFCellStyle dateCellStyle = workbook.createCellStyle();
32    dateCellStyle.setDataFormat(HSSFDataFormat.getBuiltinFormat("m/d/yy"));
33    //创建标题的显示样式
34    HSSFCellStyle headerStyle = workbook.createCellStyle();
35    headerStyle.setFillForegroundColor(IndexedColors.YELLOW.index);
36    headerStyle.setFillPattern(FillPatternType.SOLID_FOREGROUND);
37    //定义列的宽度
38    sheet.setColumnWidth(0, 5 * 256);
39    sheet.setColumnWidth(1, 12 * 256);
40    sheet.setColumnWidth(18, 20 * 256);
41    sheet.setColumnWidth(19, 12 * 256);
42    sheet.setColumnWidth(20, 8 * 256);
43    sheet.setColumnWidth(21, 25 * 256);
44    sheet.setColumnWidth(22, 14 * 256);
45    sheet.setColumnWidth(23, 12 * 256);
46    sheet.setColumnWidth(24, 12 * 256);
47    //5.设置表头
48    HSSFRow headerRow = sheet.createRow(0);
49    HSSFCell cell0 = headerRow.createCell(0);
50    cell0.setCellValue("编号");
51    cell0.setCellStyle(headerStyle);
52    HSSFCell cell1 = headerRow.createCell(1);
53    cell1.setCellValue("姓名");
54    cell1.setCellStyle(headerStyle);
55    HSSFCell cell18 = headerRow.createCell(18);
56    cell18.setCellValue("毕业院校");
57    cell18.setCellStyle(headerStyle);
58    HSSFCell cell19 = headerRow.createCell(19);
59    cell19.setCellValue("入职日期");
60    cell19.setCellStyle(headerStyle);
61    HSSFCell cell20 = headerRow.createCell(20);
62    cell20.setCellValue("在职状态");
63    cell20.setCellStyle(headerStyle);
64    HSSFCell cell21 = headerRow.createCell(21);
65    cell21.setCellValue("邮箱");
66    cell21.setCellStyle(headerStyle);
67    HSSFCell cell22 = headerRow.createCell(22);
68    cell22.setCellValue("合同期限(年)");
69    cell22.setCellStyle(headerStyle);
70    HSSFCell cell23 = headerRow.createCell(23);
```

```java
71        cell23.setCellValue("合同起始日期");
72        cell23.setCellStyle(headerStyle);
73        HSSFCell cell24 = headerRow.createCell(24);
74        cell24.setCellValue("合同终止日期");
75        cell24.setCellStyle(headerStyle);
76        //6.装数据
77        for (int i = 0; i < emps.size(); i++) {
78            HSSFRow row = sheet.createRow(i + 1);
79            Employee emp = emps.get(i);
80            row.createCell(0).setCellValue(emp.getId());
81            row.createCell(1).setCellValue(emp.getName());
82            row.createCell(18).setCellValue(emp.getSchool());
83            HSSFCell beginDateCell = row.createCell(19);
84            beginDateCell.setCellValue(emp.getBeginDate());
85            beginDateCell.setCellStyle(dateCellStyle);
86            row.createCell(20).setCellValue(emp.getWorkState());
87            row.createCell(21).setCellValue(emp.getEmail());
88            row.createCell(22).setCellValue(emp.getContractTerm());
89            HSSFCell beginContractCell = row.createCell(23);
90            beginContractCell.setCellValue(emp.getBeginContract());
91            beginContractCell.setCellStyle(dateCellStyle);
92            HSSFCell endContractCell = row.createCell(24);
93            endContractCell.setCellValue(emp.getEndContract());
94            endContractCell.setCellStyle(dateCellStyle);
95        }
96        headers = new HttpHeaders();
97        headers.setContentDispositionFormData("attachment",
98                new String("员工表.xls".getBytes("UTF-8"), "iso-8859-1"));
99        headers.setContentType(MediaType.APPLICATION_OCTET_STREAM);
100       baos = new ByteArrayOutputStream();
101       workbook.write(baos);
102   } catch (IOException e) {
103       e.printStackTrace();
104   }
105   return new ResponseEntity<byte[]>(baos.toByteArray(), headers,
106 HttpStatus.CREATED);
    }
```

代码解释：

- 首先构建一个 HSSFWorkbook 进行 Excel 基本信息配置，如文档信息、摘要信息等。
- 第 37~75 行配置列的宽度并设置表头。由于配置方式重复，因此这里省略了第 2~17 列的配置，完整配置可在 GitHub 上下载。
- 第 77~94 行表示遍历 emps 集合，将数据填充到 Excel 中。
- 第 97、98 行表示设置下载请求的文件名、编码等信息。

配置完成后，在下载请求接口中调用该方法即可，代码如下：

```java
1 @RequestMapping(value = "/exportEmp", method = RequestMethod.GET)
2 public ResponseEntity<byte[]> exportEmp() {
```

```
    return PoiUtils.exportEmp2Excel(empService.getAllEmployees());
}
```

16.8.2 前端实现

前端的实现比较简单，当用户单击“导出”按钮时，执行如下代码发起请求，下载文件：

```
window.open("/employee/basic/exportEmp", "_parent");
```

单击按钮时，会自动弹出文件保存窗口，将文件保存即可。下载后的 Excel 如图 16-20 所示。

	A	B	C	D	E	F	G	H	I	J	K	L	M	N	O
1	编号	姓名	工号	性别	出生日期	身份证号码	婚姻状况	民族	籍贯	政治面貌	电话号码	联系地址	所属部门	职称	职位
2	1	白居易	00000001	男	1990/1/1	610122199001011256	已婚	汉族	陕西	普通公民	18565558897	深圳市南山区	总办	教授	技术总监
3	2	陈静	00000002	女	1989/2/1	421288198902011234	已婚	汉族	海南	中共党员	18795556693	海南省海口市美兰区	财务部	助教	运营总监
4	3	赵琳浩	00000003	男	1993/3/4	610122199303041456	未婚	汉族	陕西	共青团员	15698887795	陕西省西安市莲湖区	技术部	助教	研发工程师
5	4	鹿存亮	00000004	男	1990/1/3	610122199001031456	已婚	汉族	陕西	共青团员	15612347795	陕西省西安市莲湖区	运维部	助教	运维工程师
6	5	姚森	00000005	男	1991/2/5	610122199102058952	已婚	汉族	河南	共青团员	14785559936	河南洛阳人民大道58号	运维部	中级工程师	运维工程师
7	6	云星	00000006	女	1993/1/5	610122199301054789	已婚	汉族	陕西西安	中共党员	15644442252	陕西西安新城区	运维部	高级工程师	运维工程师
8	7	贾旭明	00000007	男	1993/11/11	610122199311111234	已婚	汉族	广东广州	民革党员	15644441234	广东省广州市天河区冼	市场部	中级工程师	研发工程师
9	8	张黎明	00000008	男	1991/2/1	610144199102014569	已婚	汉族	广东	民建会员	18979994478	广东珠海	技术部	中级工程师	研发工程师
10	9	薛磊	00000009	男	1992/7/1	610144199207017895	已婚	汉族	陕西西安	普通公民	15648887741	西安市雁塔区	运维部	初级工程师	运维工程师
11	10	张洁	00000010	女	1990/10/9	420177199010093652	未婚	汉族	海南	民盟盟员	13695557742	海口市美兰区	运维部	高级工程师	运维工程师
12	11	白居易2	00000011	男	1990/1/1	610122199001011256	已婚	汉族	陕西	普通公民	18565558897	深圳市南山区	技术部	教授	技术总监
13	12	陈静2	00000012	女	1989/2/1	421288198902011234	已婚	汉族	海南	中共党员	18795556693	海南省海口市美兰区	华南市场部	助教	运营总监
14	13	赵琳浩2	00000013	男	1993/3/4	610122199303041456	未婚	汉族	陕西	共青团员	15698887795	陕西省西安市莲湖区	技术部	助教	研发工程师
15	14	鹿存亮2	00000014	男	1990/1/3	610122199001031456	已婚	汉族	陕西	共青团员	15612347795	陕西省西安市莲湖区	运维部	助教	运维工程师
16	15	姚森2	00000015	男	1991/2/5	610122199102058952	已婚	汉族	河南	共青团员	14785559936	河南洛阳人民大道58号	运维部	中级工程师	运维工程师
17	16	云星2	00000016	女	1993/1/5	610122199301054789	已婚	汉族	陕西西安	中共党员	15644442252	陕西西安新城区	运维部	高级工程师	运维工程师
18	17	贾旭明2	00000017	男	1993/11/11	610122199311111234	已婚	汉族	广东广州	民革党员	15644441234	广东省广州市天河区冼	市场部	中级工程师	研发工程师
19	18	王一亭	00000018	男	1991/2/1	610144199102014569	已婚	汉族	广东	民建会员	18979994478	广东珠海	技术部	中级工程师	研发工程师
20	19	薛磊2	00000019	男	1992/7/1	610144199207017895	已婚	汉族	陕西西安	普通公民	15648887741	西安市雁塔区	运维部	初级工程师	运维工程师
21	20	张洁2	00000020	女	1990/10/9	420177199010093652	未婚	汉族	海南	民盟盟员	13695557742	海口市美兰区	运维部	高级工程师	运维工程师
22	21	白居易3	00000021	男	1990/1/1	610122199001011256	已婚	汉族	陕西	普通公民	18565558897	深圳市南山区	财务部	教授	技术总监
23	22	陈静3	00000022	女	1989/2/1	421288198902011234	已婚	汉族	海南	中共党员	18795556693	海南省海口市美兰区	华南市场部	助教	运营总监
24	24	鹿存亮3	00000024	男	1990/1/3	610122199001031456	已婚	汉族	陕西	共青团员	15612347795	陕西省西安市莲湖区	运维部	助教	运维工程师
25	25	姚森3	00000025	男	1991/2/5	610122199102058952	已婚	汉族	河南	共青团员	14785559936	河南洛阳人民大道58号	运维部	中级工程师	运维工程师

图 16-20

经过如上配置后，员工数据导出功能就实现了，完整代码读者可以参考 https://github.com/lenve/vhr。

16.9 员工资料导入

既然有员工资料导出需求，当然也就有导入需求。对前端而言，员工资料导入就是文件上传，对后端而言，则是获取上传的文件进行解析，并把解析出来的数据保存到数据库中。

16.9.1 后端接口实现

后端主要是获取前端上传文件的流，然后进行解析，代码如下：

```
public static List<Employee> importEmp2List(MultipartFile file,
                                            List<Nation> allNations,
                                            List<PoliticsStatus> allPolitics,
                                            List<Department> allDeps,
                                            List<Position> allPos,
                                            List<JobLevel> allJobLevels) {
    List<Employee> emps = new ArrayList<>();
    try {
```

```java
 9      HSSFWorkbook workbook =
10              new HSSFWorkbook(new POIFSFileSystem(file.getInputStream()));
11      int numberOfSheets = workbook.getNumberOfSheets();
12      for (int i = 0; i < numberOfSheets; i++) {
13      HSSFSheet sheet = workbook.getSheetAt(i);
14      int physicalNumberOfRows = sheet.getPhysicalNumberOfRows();
15      Employee employee;
16      for (int j = 0; j < physicalNumberOfRows; j++) {
17      if (j == 0) {
18          continue;//标题行
19      }
20      HSSFRow row = sheet.getRow(j);
21      if (row == null) {
22          continue;//没数据
23      }
24      int physicalNumberOfCells = row.getPhysicalNumberOfCells();
25      employee = new Employee();
26      for (int k = 0; k < physicalNumberOfCells; k++) {
27      HSSFCell cell = row.getCell(k);
28      switch (cell.getCellTypeEnum()) {
29      case STRING: {
30          String cellValue = cell.getStringCellValue();
31          if (cellValue == null) {
32              cellValue = "";
33          }
34      switch (k) {
35          case 1:
36              employee.setName(cellValue);
37              break;
38          case 2:
39              employee.setWorkID(cellValue);
40              break;
41          case 3:
42              employee.setGender(cellValue);
43              break;
44          case 5:
45              employee.setIdCard(cellValue);
46              break;
47          case 6:
48              employee.setWedlock(cellValue);
49              break;
50          case 7:
51              int nationIndex = allNations.indexOf(new Nation(cellValue));
52              employee.setNationId(allNations.get(nationIndex).getId());
53              break;
54          case 8:
55              employee.setNativePlace(cellValue);
56              break;
57          case 9:
58              int psIndex = allPolitics.indexOf(new PoliticsStatus(cellValue));
```

```java
                employee.setPoliticId(allPolitics.get(psIndex).getId());
                break;
            case 10:
                employee.setPhone(cellValue);
                break;
            case 11:
                employee.setAddress(cellValue);
                break;
            case 12:
                int depIndex = allDeps.indexOf(new Department(cellValue));
                employee.setDepartmentId(allDeps.get(depIndex).getId());
                break;
            case 13:
                int jlIndex = allJobLevels.indexOf(new JobLevel(cellValue));
                employee.setJobLevelId(allJobLevels.get(jlIndex).getId());
                break;
            case 14:
                int posIndex = allPos.indexOf(new Position(cellValue));
                employee.setPosId(allPos.get(posIndex).getId());
                break;
            case 15:
                employee.setEngageForm(cellValue);
                break;
            case 16:
                employee.setTiptopDegree(cellValue);
                break;
            case 17:
                employee.setSpecialty(cellValue);
                break;
            case 18:
                employee.setSchool(cellValue);
                break;
            case 19:
            case 20:
                employee.setWorkState(cellValue);
                break;
            case 21:
                employee.setEmail(cellValue);
                break;
        }
    }
    break;
    default: {
        switch (k) {
            case 4:
                employee.setBirthday(cell.getDateCellValue());
                break;
            case 19:
                employee.setBeginDate(cell.getDateCellValue());
                break;
```

```
109                 case 22:
110                     employee.setContractTerm(cell.getNumericCellValue());
111                     break;
112                 case 23:
113                     employee.setBeginContract(cell.getDateCellValue());
114                     break;
115                 case 24:
116                     employee.setEndContract(cell.getDateCellValue());
117                     break;
118             }
119         }
120         break;
121     }
122     }
123     emps.add(employee);
124     }
125     }
126     } catch (IOException e) {
127         e.printStackTrace();
128     }
129     return emps;
130 }
```

代码解释：

- 首先根据上传文件的流获取一个 HSSFWorkbook 对象，然后获取 workbook 中表单的个数，进行遍历。
- 对于每一个表单，首先获取行数，然后进行遍历，第一行是标题行，跳过，如果该行没有数据也跳过，如果该行数据正常，就获取该行的单元格个数进行遍历。
- 本案例中，单元格的格式主要分为三种，即日期、数字以及普通文本，因此在不同的 switch 分支中进行处理。
- 最后将遍历得到的员工数据集合返回。

在数据导入接口中调用 importEmp2List 方法，获取员工数据集合后，插入数据库中即可，代码如下：

```
1  @RequestMapping(value = "/importEmp", method = RequestMethod.POST)
2  public RespBean importEmp(MultipartFile file) {
3      List<Employee> emps = PoiUtils.importEmp2List(file,
4              empService.getAllNations(), empService.getAllPolitics(),
5              departmentService.getAllDeps(), positionService.getAllPos(),
6              jobLevelService.getAllJobLevels());
7      if (empService.addEmps(emps) == emps.size()) {
8          return RespBean.ok("导入成功!");
9      }
10     return RespBean.error("导入失败!");
11 }
```

16.9.2 前端实现

前端主要是一个 Excel 表格的上传，这里直接采用 Element 的文件上传控件，代码如下：

```html
<el-upload
  :show-file-list="false"
  accept="application/vnd.ms-excel"
  action="/employee/basic/importEmp"
  :on-success="fileUploadSuccess"
  :on-error="fileUploadError"
  :disabled="fileUploadBtnText=='正在导入'"
  :before-upload="beforeFileUpload"
  style="display: inline">
<el-button size="mini" type="success"
  :loading="fileUploadBtnText=='正在导入'">
<i class="fa fa-lg fa-level-up"></i>
  {{fileUploadBtnText}}
</el-button>
</el-upload>
```

代码解释：

- accept 表示接收的上传文件类型。
- action 表示上传接口。
- :on-success 表示上传成功时的回调。
- :on-error 表示上传失败时的回调。
- :disabled 表示当 fileUploadBtnText 属性的值为"正在导入"时禁用上传控件。这个配置主要考虑到上传是一个耗时操作，在一个文件上传的过程中，其他文件暂时不能上传。
- :before-upload 表示文件上传前的回调。
- el-button 中的:loading="fileUploadBtnText=='正在导入'"表示当 fileUploadBtnText 的文本为"正在导入"时，显示一个 Loading 加载。

相关回调方法如下：

```js
fileUploadSuccess(response, file, fileList){
  if (response) {
    this.$message({type: response.status, message: response.msg});
  }
  this.loadEmps();
  this.fileUploadBtnText = '导入数据';
},
fileUploadError(err, file, fileList){
  this.$message({type: 'error', message: "导入失败!"});
  this.fileUploadBtnText = '导入数据';
},
beforeFileUpload(file){
  this.fileUploadBtnText = '正在导入';
}
```

代码解释：

- 在文件上传之前，首先设置 fileUploadBtnText 的文本为"正在导入"，这样上传按钮上的文本就会变为"正在导入"，同时上传按钮的状态变为禁用，并且在上传按钮上多了一个 Loading。
- 在上传成功时，给用户以提示，然后重新加载员工数据，并将 fileUploadBtnText 的文本设置为"导入数据"。
- 上传出错时，给用户以提示，同时将 fileUploadBtnText 的文本设置为"导入数据"。

所有配置完成后，单击"导入数据"，选择 16.8 节导出的用户数据进行导入，如图 16-21 所示。

图 16-21

16.10　在线聊天

在线聊天是一个为了方便 HR 进行快速沟通提高工作效率而开发的功能,考虑到一个公司中的 HR 并不多，并发量不大，因此这里直接使用最基本的 WebSocket 来完成该功能。

16.10.1　后端接口实现

要使用 WebSocket，首先引入 WebSocket 依赖：

```
1  <dependency>
2  <groupId>org.springframework.boot</groupId>
3  <artifactId>spring-boot-starter-websocket</artifactId>
4  </dependency>
```

依赖添加成功后，接下来配置 WebSocket 配置类，代码如下：

```
1  @Configuration
2  @EnableWebSocketMessageBroker
3  public class WebSocketConfig extends AbstractWebSocketMessageBrokerConfigurer {
4      @Override
5      public void registerStompEndpoints(StompEndpointRegistry stompEndpointRegistry)
6  {
7          stompEndpointRegistry.addEndpoint("/ws/endpointChat").withSockJS();
8      }
```

```
9
10      @Override
11      public void configureMessageBroker(MessageBrokerRegistry registry) {
12          registry.enableSimpleBroker("/queue","/topic");
13      }
14  }
```

然后创建消息转发 Controller，代码如下：

```
1   @Controller
2   public class WsController {
3       @Autowired
4       SimpMessagingTemplate messagingTemplate;
5
6       @MessageMapping("/ws/chat")
7       public void handleChat(Principal principal, String msg) {
8           String destUser = msg.substring(msg.lastIndexOf(";") + 1, msg.length());
9           String message = msg.substring(0, msg.lastIndexOf(";"));
10          messagingTemplate.convertAndSendToUser(destUser, "/queue/chat", new
11  ChatResp(message, principal.getName()));
12      }
13      @MessageMapping("/ws/nf")
14      @SendTo("/topic/nf")
15      public String handleNF() {
16          return "系统消息";
17      }
18  }
```

配置完成后，重启后端项目，然后开始配置前端。

16.10.2 前端实现

聊天功能写在 FriendChat.vue 组件中，但是用户登录成功后，首先加载的是 Home.vue 页面，在 Home 页面的右上角有一个通知的图标，如果有最新的通知，这里会显示一个红点，如图 16-22 所示。

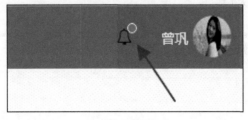

图 16-22

因此，虽然聊天是在 FriendChat.vue 页面进行的，但是 WebSocket 连接却需要登录成功后才建立，这里选择在 store 中建立 WebSocket 请求，代码如下：

```
1   import Vue from 'vue'
2   import Vuex from 'vuex'
```

```js
import '../lib/sockjs'
import '../lib/stomp'

Vue.use(Vuex)

export default new Vuex.Store({
  state: {
    routes: [],
    msgList: [],
    isDotMap: new Map(),
    currentFriend: {},
    stomp: Stomp.over(new SockJS("/ws/endpointChat")),
    nfDot: false
  },
  mutations: {
    toggleNFDot(state, newValue){
      state.nfDot = newValue;
    },
    updateMsgList(state, newMsgList){
      state.msgList = newMsgList;
    },
    updateCurrentFriend(state, newFriend){
      state.currentFriend = newFriend;
    },
    addValue2DotMap(state, key){
      state.isDotMap.set(key, "您有未读消息")
    },
    removeValueDotMap(state, key){
      state.isDotMap.delete(key);
    }
  },
  actions: {
    connect(context){
      context.state.stomp = Stomp.over(new SockJS("/ws/endpointChat"));
      context.state.stomp.connect({}, frame=> {
        context.state.stomp.subscribe("/user/queue/chat", message=> {
          //接收在线聊天消息
        });
        context.state.stomp.subscribe("/topic/nf", message=> {
          //接收系统通知
        });
      }, failedMsg=> {

      });
    }
  }
});
```

定义好之后,在初始化菜单数据的地方调用 connect 方法建立 WebSocket 连接,代码如下:

```js
export const initMenu = (router, store)=> {
```

```
2    if (store.state.routes.length > 0) {
3      return;
4    }
5    getRequest("/config/sysmenu").then(resp=> {
6      if (resp && resp.status == 200) {
7        var fmtRoutes = formatRoutes(resp.data);
8        router.addRoutes(fmtRoutes);
9        store.commit('initMenu', fmtRoutes);
10       store.dispatch('connect');
11     }
12   })
13 }
```

通过 store.dispatch('connect');调用 connect 方法。

这里配置完成后，重新登录，在 Chrome 控制台可以看到 WebSocket 连接已经成功建立起来了，如图 16-23 所示。

```
Opening Web Socket...                              stomp.js?195a:145
Web Socket Opened...                               stomp.js?195a:145
>>> CONNECT                                        stomp.js?195a:145
accept-version:1.1,1.0
heart-beat:10000,10000


<<< CONNECTED                                      stomp.js?195a:145
version:1.1
heart-beat:0,0
user-name:admin


connected to server undefined                      stomp.js?195a:145
>>> SUBSCRIBE                                      stomp.js?195a:145
id:sub-0
destination:/user/queue/chat


>>> SUBSCRIBE                                      stomp.js?195a:145
id:sub-1
destination:/topic/nf
```

图 16-23

最后，在 FriendChat.vue 中通过如下方式发送一条消息：

```
1  this.$store.state.stomp.send("/ws/chat", {}, this.msg + ";" +
   this.currentFriend.username);
```

另外，浏览器在收到消息之后，是将消息保存在 store 中的，这样一旦收到消息，FriendChat 页面的聊天数据就会自动更新，并且，当有新消息到达时，即使用户不在 FriendChat 页面，也能及时收到通知（主页右上角的通知图标会显示小红点）。

聊天效果如图 16-24、图 16-25 所示。

图 16-24

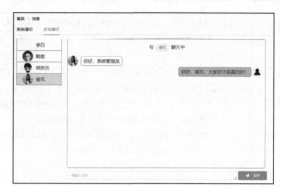

图 16-25

这里由于前端页面代码量庞大，因此只贴出部分关键步骤的代码，完整代码读者可以在 https://github.com/lenve/vhr 下载。

这里有两个订阅，"/user/queue/chat" 是用来做在线聊天的，"/topic/nf" 则是为了接收系统通知。

16.11 前端项目打包

当前端项目开发完成后，执行如下命令对项目进行打包：

```
1  npm run build
```

执行结果如图 16-26 所示。

图 16-26

打包完成后，在当前工作目录下生成一个 dist 文件夹，如图 16-27 所示。将里边的 index.html

和 static 目录复制到 Spring Boot 项目的 static 目录下，如图 16-28 所示。

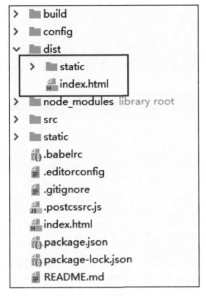

图 16-27

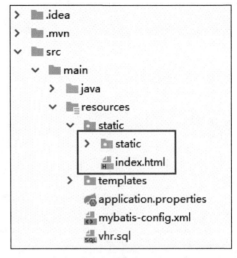

图 16-28

接下来，启动后端项目，直接在浏览器中输入 http://localhost:8082/index.html 就可以看到登录页面，如图 16-29 所示。此时就可以将该 Spring Boot 项目直接打包发布（参见第 15 章）。

图 16-29

16.12 小　　结

本章向读者介绍了一个微人事项目，主要从登录模块、动态加载用户菜单、员工资料模块、邮件发送模块、Excel 导入导出模块、在线聊天模块以及编译打包几个方面介绍。由于原项目代码量庞大，本章主要选取一些关键步骤进行介绍，完整代码读者可以在 GitHub 上下载，下载地址为 https://github.com/lenve/vhr。

16.12 小 结

图 16-29

图 16-30